81 8710193 1
TELEPEN
WITHDRAWN FROM
AF848723

# Biomedical Institutions, Biomedical Funding, and Public Policy

CPC UCW
LLYFRGELL
LIBRARY
ABERYSTWYTH

# Biomedical Institutions, Biomedical Funding, and Public Policy

Edited by
H. Hugh Fudenberg
*Medical University of South Carolina*
*Charleston, South Carolina*

PLENUM PRESS · NEW YORK AND LONDON

Library of Congress Cataloging in Publication Data

Main entry under title:

Biomedical institutions, biomedical funding, and public policy.

Includes bibliographical references and index.

1. Medical research—United States. 2. Medical research—United States—Cost effectiveness. 3. Medical policy—United States. 4. Genetic engineering. 5. Hybridomas. I. Fudenberg, H. Hugh. [DNLM: 1. Economics, medical—Trends—United States. 2. Health policy—United States. 3. Research—United States. 4. Research support—Trends. W 20.5 B6164]

R854.U5B517 1983 610'.72073 83-11154

ISBN 0-306-41231-4

Prifysgol Cymru
rsity College Of Wales
berystwyth
8187101931
R854 U5B6

© 1983 Plenum Press, New York
A Division of Plenum Publishing Corporation
233 Spring Street, New York, N.Y. 10013

All rights reserved

No part of this book may be reproduced, stored in a retrieval system, or transmitted in any form or by any means, electronic, mechanical, photocopying, microfilming, recording, or otherwise, without written permission from the Publisher

Printed in the United States of America

Senator Charles McC. Mathias, Jr. (left) and Dr. H. Hugh Fudenberg (right).

# Contributors

**Susan Bennett** • The Upjohn Company, Kalamazoo, Michigan 49001

**Graham V. Brown** • The Walter and Eliza Hall Institute of Medical Research, Royal Melbourne Hospital, Victoria 3050, Australia

**Leighton E. Cluff** • The Robert Wood Johnson Foundation, Princeton, New Jersey 08540

**Theodore Cooper** • The Upjohn Company, Kalamazoo, Michigan 49001

**Jean Dausset** • Université de Paris VII, Institut de Recherches sur les Maladies du Sang, Hôpital Saint-Louis, 75475 Paris, France

**Burton C. D'Lugoff** • Department of Medicine, Baltimore City Hospital and Johns Hopkins University School of Medicine, Baltimore, Maryland 21224

**Victor R. Fuchs** • Department of Economics, Stanford University, and National Bureau of Economic Research, Stanford, California 94305

**H. Hugh Fudenberg** • Department of Basic and Clinical Immunology and Microbiology, Medical University of South Carolina, Charleston, South Carolina 29425

**J. Leslie Glick** • Genex Corporation, Rockville, Maryland 20852

**H. C. Goodman** • Tropical Medicine Center, Johns Hopkins University, Baltimore, Maryland 21205

**Yuet Wai Kan** • Howard Hughes Medical Institute Laboratory and Department of Medicine, University of California at San Francisco, San Francisco, California 94143

**T. A. Lambo** • World Health Organization, Geneva, Switzerland

**Hans Löw** • Department of Endocrinology, Karolinska Hospital, Stockholm, Sweden

**Rolf Luft** • Department of Endocrinology, Karolinska Hospital, Stockholm, Sweden

**Hon. Charles McC. Mathias, Jr.** • United States Senator from Maryland, Washington, D.C. 20510

**G. J. V. Nossal** • The Walter and Eliza Hall Institute of Medical Research, Royal Melbourne Hospital, Victoria 3050, Australia

**Felix T. Rapaport** • Transplantation Service, Department of Surgery, State University of New York at Stony Brook, Stony Brook, New York 11794

**David E. Rogers** • The Robert Wood Johnson Foundation, Princeton, New Jersey 08540

**Chester W. Schmidt, Jr.** • Department of Psychiatry, Baltimore City Hospital and Johns Hopkins University School of Medicine, Baltimore, Maryland 21224

**Philip D. Zieve** • Department of Medicine, Baltimore City Hospital and Johns Hopkins University School of Medicine, Baltimore, Maryland 21224

# Preface

The world is on the threshold of a great new industrial revolution, a scientific–industrial revolution.[1] Recombinant DNA technology and hybridoma technology ("monoclonal antibodies") have already provided unique investment opportunities for venture capitalists. Hence published reports of biomedical research are no longer restricted to scientific journals, but now appear regularly not only in weekly newsmagazines like *Time*[2] and *U.S. News & World Report*,[3] but also in the financial sections of *The New York Times*,[4] *The Wall Street Journal*,[5] *Business Week*,[6] *Fortune*,[7] and *The Economist*,[8] as well as in such stock market advisory publications as *New Issues*[9] and *Inc. (The Magazine for Growing Companies)*.[10] These publications now appear to be as important to biomedical scientists in keeping abreast of new scientific developments in biotechnology as is *Current Contents*. (The costs of health-cost provision and of fundamental biomedical research are now also being followed by such media.) Conversely, Wall Street financial brokers increasingly no longer confine their reading to economic journals but are also perusing *Nature*,[11] *Science*,[12] and *Science News*[13] for information on both fiscal and scientific advances in these areas. It is obvious that the information explosion in biotechnology is crossing traditional boundaries (e.g., ref. 14).

This volume is the second of several that are intended to inform both the biomedical community and interested intelligent laymen of the political and economic implications of biomedical research. The first volume, *Biomedical Scientists and Public Policy* (1978), covered many areas: the chapter on fiscal benefits of biomedical research indicated that the cost–benefit ratio of such research appeared to be at least $100 saved for every $1 spent for basic biomedical research. These data were instrumental in helping the Canadian Medical Research Council to obtain a 30% increase in funding for biomedical research throughout Canada[15] and in enabling the Alberta Heritage Research Foundation[16] to generate marked increases in the foundation's grants in this area. The current volume focuses interest in three areas:

1. The role of foundations in the formulation of biomedical research policy and the evaluation of similarities and differences in various biomedical research establishments (e.g., organized academic research institutions, the NIH, foundations, and the pharmaceutical industry).
2. Analyses of both the costs of fundamental research within a given field and the subsequent economic "fallout."
3. An update on "dollar savings" due to basic biomedical research, past and future (estimated).

Nowhere are the probable economic benefits so great as in recombinant DNA technology ("genetic engineering"), which evolved from molecular biology, the study of the molecular basis of inheritance. Recombinant DNA is DNA characterized by insertion of foreign genetic material into a living host (usually the *Escherichia coli* bacterium) capable of producing, or "expressing," proteins specified by the genetic material which has been inserted. This means that many proteins of potential significance in such areas as medicine, agriculture, animal husbandry, and, further into the future, energy technology, can be manufactured in almost unbounded quantities. (The economics of this process ultimately will prove little more costly than that of brewing technology.[17])

Ten years ago molecular biology concentrated almost solely on the genetics of a lambda bacteriophage (a virus) confined to a strain of *E. coli* (a nonpathogenic bacterium which normally inhabits the human intestinal tract) and was considered totally devoid of "relevance"; indeed, molecular biology was considered to be more of an ivory tower than any of the various other research areas in biology. Insofar as we can ascertain (see Chapter 3), total NIH funding devoted to basic research in this area until 1974 totaled $91 million. It is estimated that we will save $3–5 billion per year as a result of this $91 million by the early 1990s.

Not only has recombinant DNA already provided an approach to therapy in certain human disorders, but in addition molecular biology has resulted in new methods for diagnosis of herditary disorders, including antenatal diagnosis (see Chapter 4). Consider a yet-unborn child doomed to a considerably shortened life-span and severe intractable pain owing to any one of the many genetic disorders potentially detectable *in utero,* ranging from adenosine deaminase deficiency to xeroderma pigmentosum.[18] The cost savings will be enormous should the child's parents elect abortion in the first trimester rather than a few years of institutionalization. Furthermore, antenatal diagnosis which turns out normal will spare the parents of a yet-unborn child the emotional anguish of uncertainty during the last two trimesters. (How can one estimate the economic value of this increased peace of mind and hence productivity?)

Savings of similar magnitudes appear possible through another microbiologic discovery—hybridoma formation. The use of highly specific antibodies manufactured by this method holds great promise. Not only will such antibodies be useful in diagnostic tests for a wide variety of antigens including bacteria and viruses, but they appear to offer special promise in cancer diagnosis, cancer prognosis, and cancer immunotherapy[19] (targeted antibodies, like the proverbial "silver bullets," will come to cancerous tissues bearing tumor-specific antigens[20]).

Similarly, leading journals[21] discuss the billions of dollars to be saved through biotechnology. Thus the economist and the scientist are beginning to derive their information from overlapping sources; furthermore the interactions of informed scientists with legislators on the state and federal levels, with health economists, with science writers, and with intelligent laymen, are increasingly helping to formulate public policy with respect to both science and health. In the course of such interactions, at least some scientists have become exposed to points of view not previously encountered, and they are aiding, especially in the biological sciences, to shape national priorities and goals. Unfortunately most of the published information necessary to form opinions in these areas is available only in widely scattered publications encompassing many disciplines (as exemplified by the spectrum mentioned previously) and in daily and weekly newspapers, weekly and monthly journals (both medical and economic), and yearly *Advances* volumes.

This volume is designed to place between two covers sufficient information concerning various aspects of biomedical institutions, biomedical funding, and public policy to induce both intelligent laymen and biomedical scientists to venture further in garnering economic information in this field. It is not easily understandable why biomedical scientists have failed to use such data in the past. I hope that this volume will make it easier for them to do so in the future.

H. Hugh Fudenberg

## REFERENCES


1. *Biotechnology and the Stock Market,* 1980, The Institute for Econometric Research, *Market Logic,* Fort Lauderdale, Florida, June 6, pp. 1–4.
2. Faded genes: Slow start for bioengineering, April 19, 1982, . *Time.*
3. *Research gets back to basic on human cells,* 1980 *U.S. News & World Report,* September 15, p. 55.
4. Molecular genetics, focusing on agriculture, to go public, 1982, *New York Times,* April 27, p. 36.
5. Bishop, J. E., 1981, Armed antibodies: Researchers are closer to biological "Missles," *Wall Street Journal,* June 3; Abbott Labs is cleared by FDA to use in tests monoclonal antibodies, 1982, *Wall Street Journal,* April 14, p. 12.
6. A potential cool against cancer (monoclonal antibodies), 1981, *Business Week,* May 18, p. 154.
7. Alexander, T., Teaching computers the art of reason, 1982, *Fortune,* May 17, pp. 82–92.
8. Biotechnology: Mini monos, 1981, *The Economist,* November 14, p. 109.
9. Fosback, N. G. (ed.), 1982 *Molecular Genetics: New Issues,* Glen King Parker, Publisher, Fort Lauderdale, pp. 1–2.
10. Gene transfers may build a better cow, 1982, *Inc.,* May, p. 21.
11. Mixed welcome for genes on Wall Street: Genentech's interferon from yeast plasmid, 1981, *Nature,* **290**:77; Recombinant DNA guidelines wear thin: NIH moves to voluntary regulation, 1981, *Nature,* **291**:3.
12. Pajaro Dunes: The search for consensus, 1982, *Science* **216**:155.
13. Interferon by bacteria, on trial, 1982, *Science News* **119**:54.
14. Anderson, W. F., French, D., and Elaine, G., Genetic engineering in mammalian cells, 1981, *Sci. Am.* **245**:106.

15. Gold, P., McGill University, Montreal, Canada (personal communication), 1982.
16. Bradley, J., Executive Director, Alberta Heritage Foundation, Edmonton, Canada (personal communication), 1982.
17. Tenth Annual Hambrecht & Quist Technology Conference, April 26–29, 1982, San Francisco, California.
18. Golbus, M. S. The current scope of antenatal diagnosis, 1982, *Hospital Practice* **17:**179–180.
19. Bernstein, I. D., Nowinski, R. C., and Wright, P. W., 1982, Monoclonal antibody diagnosis and treatment of malignant disease *in: Monoclonal Antibodies; Prospects for Cancer Treatment, Immunologic Aspects of Cancer Therapeutics,* pp. 35–41.
20. Marx, J. L., Monoclonal antibodies in cancer, 1982, *Science* **216:**283–285.
21. Thatcher applauds genes, 1981, *Nature* **290.**

# Acknowledgments

This volume was made possible by the interest and cooperation not only of those who contributed its chapters, but also of many concerned scientists, legislators, and others with whom we have spent untold hours in garnering data for evaluation. I am also grateful to Dr. H. D. Whitten for his invaluable editorial assistance and to H. Rhett Coombs Halverson for her assistance in obtaining data.

H. H. F.

# Contents

# Introduction

CHARLES McC. MATHIAS, Jr.

This book focuses on three areas which lie at the heart of biomedical science in the United States—biomedical institutions, biomedical funding, and public policy. In a sense they are the "heart valves" of American science and they all depend to varying degrees on government support. They are all therefore susceptible to public influence.

The ability of science to serve society to the fullest extent possible and in a way that is responsive to society's goals depends on the health and efficiency of these "valves" and on how well they work together. The malfunction of any one of them strikes at the heart of science and of course negatively affects the body politic. To assure that all function smoothly together, the scientific community must become more actively involved in the formulation of public policy on health and science issues, and policy-makers and the public must become more informed about the needs of the scientific community and the problems it confronts. This volume, written and edited by scientists, should act as a powerful catalyst to that necessary interaction.

It is surely not coincidental that this book comes at a time when the public is increasingly interested in and concerned about science policy issues. In recent years biomedical science, its consequences, implications, and potential impact on our society, have attracted greater public attention and scrutiny than at any time in the last half century. Increased public interest in various facets of biomedical science is reflected in the upsurge of science coverage by the news

---

HON. CHARLES McC. MATHIAS, Jr. • United States Senator from Maryland, Washington, D.C. 20510.

media, in the popularity of science magazines and television programs on science designed to attract scientifically unsophisticated viewers. But perhaps the most significant reflection of public interest in biomedical science can be found in the activities of the Congress which address biomedical science issues.

Any sampling of recent newspaper headlines would demonstrate clearly that the three areas highlighted in this book are topics of broad public concern. The prominent placement of science news articles is striking. A decade ago biomedical science news, if it made the papers at all, was buried in the back pages. Today more often than not it is front-page news. For example, not long ago *The Washington Post* ran a front-page series on genetic engineering and recombinant DNA technology which provided a comprehensive review of the highly significant scientific advances made in this field in recent years. *Baltimore Magazine,* a popular Maryland monthly, recently enlightened its readers about the techniques involved in this new research in an entertaining article entitled "A Blueprint for Shaping Life: How To Make Your Own Clone." Other examples are legion.

Scientists have long sought such media coverage, but now that it has come about, it imposes new burdens on the scientific community. When an issue commands broad public attention, it almost inevitably also commands the attention of those who make policy. When that happens, legislative or regulatory action often ensues, resulting in new or modified public policies. Given this type of chain reaction, it behooves the science community to keep a finger on the public pulse in order to anticipate areas of concern and to work cooperatively with the public to shape opinion in those areas. The public interest requires scientists to justify a strong national commitment to scientific work and to foster public understanding of the profound significance that new directions in science have for our society and our daily lives.

In the first two years of the Reagan administration important public policy questions were framed almost exclusively in economic terms, and the issues of biomedical science, like many others, were debated in a budgetary context. The philosophy of the federal government's involvement in almost every sector of national life was reexamined with a fine-tooth comb and the bottom line has been cuts more or less across the board. In my view, funding cuts for basic research are short-sighted. But scientists and policy-makers must face the fact that there is broad public support for holding down the federal budget and that areas which have attracted public criticism are almost certain to come under the knife. The rare but well-publicized instances of fraud in biomedical research and misuse of funds obviously do not enhance chances for biomedical research funding. Here again it falls to policy-makers and scientific administrators, working in concert, to ensure that research dollars are applied effectively and in a way that will inspire public confidence and public support for the substantial funding for scientific training and research that the times demand.

Uncertainty about funding and unanswered questions about policy pose serious problems for our biomedical institutions. When budget reductions in science education were announced, much was said and written about the delapidated state of the science departments in America's colleges and universities. Outdated equipment is a grim reality. Faculty and laboratory shortages in some science fields persist because of academia's inability to compete with salaries and benefits offered by private industry. This situation will worsen as advances in the genetic engineering industries lead to increased competition for the best scientific minds.

These conditions have intensified the debate on the government's role in our academic and research institutions. Support for scientific research through increased industry–academia cooperation is another variable. Some good ideas have been discussed for years, and now finally are beginning to be applied. But it would be a great mistake to believe that industry alone can provide an immediate or comprehensive solution. Over the last decade the federal government funded 60% of the nation's health research and development, with the next largest contribution coming from the pharmaceutical industry. I doubt that industry—with its need to show a profit in the short term—will assume this task. The commercial payoffs are simply too far away.

Therefore, even in these days of severely strained federal budgets, it would be the worst kind of fiscal improvidence to decrease significantly our national financial commitment to basic biomedical research, especially in view of the considerable fiscal benefits we can expect to derive from such research in the long run (see Chapter 3).

Academic institutions, scientists, and industry officials must recognize the potential of cooperation and investment to strengthen research activity. In recognition of budgetary constraints and the national debate, our biomedical institutions must not only adapt to change, but they must initiate it. This is especially critical in the areas of funding and training. Institutions and scientists must continue to look for new ways to fund research and training and to upgrade laboratory facilities as part of their mission.

But there is a vital fact to remember: Institutions do not and perhaps cannot adapt overnight. Government has a responsibility to recognize this fact and to lend its assistance. This is one area of the budget where any limitations should be worked out delicately and with surgical instruments, not by hacking away with a butcher's cleaver. Programs that do not produce an obvious result or serve an immediate and specific need are particularly vulnerable. The scientific method is slow and painstaking, but the rewards are real and substantial—alleviation of suffering, improved health, reduced health-care costs, and increased productivity by our citizens.

Winston Churchill observed that "Modern science is standing on tiptoe, ready to open the doors of a golden age." There is no doubt that developments in biomedical science have pushed those doors open. But whether or not we will actually enter the "golden age" Churchill foresaw depends very largely on what happens to the institutions, funding, and public policy this book examines.

Unlike the space program of the 1960s, biomedical science does not have the competition of a "Sputnik" to rally the American people behind it. Every detail of space exploration from the first earth orbit through the moon landings and the Jupiter probe has been reported in proud and loving detail in our media. The American people have seen these miracles with their own eyes.

The miracles of biomedical science are every bit as real, and sometimes just as visible, but the reaction to them is quite different. The first public reaction to recombinant DNA research and other types of genetic manipulation was a crescendo of concern about the danger and potential abuses of such research. The scientific community is responding by increasing its emphasis on public education, but these fears will not be dispelled overnight. In the meantime scientists must keep the lines of communication open, while adequate systems of monitoring ensure that research is conducted within policies and guidelines.

It is particularly important that government, industry, and academia work together now to strengthen American science and our network of scientists and research institutions. The public's increased interest in biomedical science, even its apprehensions, should certainly serve to promote responsibility and sensitivity in the scientific community and among those responsible for the formulation of public policy.

If the scientific community expects the public to pay the bills, it should assume a significant part of the burden of amassing, evaluating, and communicating the facts to justify the commitment of public funds. This volume is a major contribution to that effort, but the effort must be continuous.

# 1

# Excellence and Creativity in Science

ROLF LUFT and HANS LÖW

The great discoveries within the natural sciences and medicine are usually followed by explosive development in the new area: an explorative and exploitive phase, where the original findings are reproduced and filled out to be incorporated into our thinking. At its best this leads to a widening of the framework of ideas and enables the completion of partly formulated ideas. During this exploitive phase of research the quality of *excellence* plays a decisive role. Excellent or outstanding are those who grasp rapidly the distinctiveness of a new concept, those who are endowed with knowledge wide enough to exploit fully the new concept and who in addition have the necessary technical skill. However, the one who created the opportunities for this development was the one who gave birth to the new idea—the innovator—*the creator.*

The concepts of excellence and creativity are not limited to science but apply to all the varied activities in society. We also meet these concepts in art. Over long periods artists express thoughts, feelings, and impressions in a way that may be considered traditional for the time. In the middle of this phase, suddenly there comes forward a pioneer, an innovator, who introduces a new interpretation of the reality he wants to describe. Later, many will follow who begin to employ and further develop this new style and technique—many even with a masterly touch.

Let us examine two examples. *The Annunciation,* painted by Simone Martini in 1333, is shown in Figure 1. It hangs in the Uffizi in

ROLF LUFT and HANS LÖW • Department of Endocrinology, Karolinska Hospital, Stockholm, Sweden. The text of this chapter was first published in *Clinical Research,* 1980, **4**:329–333.



U.C.W. ABERYSTWYTH

**Figure 1.** *The Annuciation* by Simone Martini, 1333.

Florence. This mannered gothic style that prevailed—indeed blossomed, for flowers were the favorite motif, over all of Europe during the 14th century—really had the character of excellence: a highly developed craftmanship and artistry. Simone Martini's *Annunciation* is a fine example: well mannered, graceful, and restrained.

At this time Masaccio made his entrance in Florence and completely abandoned this refined craftmanship for a new creative concept (Figure 2). In his fresco *The Holy Trinity with the Virgin and St. John* (1425), in the Santa Maria Novella church in Florence, he abandoned all the beauty for one end: the overwhelming image of space, built upon one single creative innovation, the central perspective. This obviously simple but innovative geometrical idea is even today the basis for one of the commonest of all instruments, the camera.

Excellence and creativity have always occurred side-by-side in science, but their relative weights have varied. At the turn of the century the conditions for excellent and creative research workers

**Figure 2.** *The Holy Trinity with the Virgin and St. John* by Masaccio, 1425.

were different from those today. First, both the world of knowledge and the world of concepts were more limited; therefore it was easier to achieve the necessary standard of excellence in a given field. Second, the attainments in science were attributed most often to a single person. His achievements were easier to recognize since the breadth of scientific information was so limited. Third, the demand for technical resources was so much less. Even if the available economic resources were small, they were seldom a limiting factor. Indeed, the researcher was often economically self-supporting; research was carried on as a spare-time occupation or even by the dilettante. These conditions fostered creativity as well as excellence.

The situation today is quite different. First, the world of knowledge has become almost overwhelming in its complexity. Even the acquisition of basic knowledge makes special demands on the individual's receptivity and ability to be selective. Second, and partly as a result of this complexity, today's research is seldom a "one-man show," but rather the result of collaboration between different types of individuals and between specialists in different fields. Collaboration is the rule, if not an art. Third, the demand for technical skills and technical resources dominates. For this reason by far the greatest part of the resources allocated to research is used for salaries to the people with the required technical knowledge—to buy the commodity of excellence.

In this ever more complicated conceptual world, excellence has become dominant. One must possess considerable excellence even to be able to pose the relevant questions. We even have an excess of excellence to employ today. It is now the economic restrictions that set the limits, making it necessary to give priority and support only to research projects that can be expected to give especially worthwhile results.

The increasing need to obtain funds before research can proceed has had a profound effect on the possibilities for creativity and excellence today. By and large during the last 25 years research grants have been allocated to projects and not, as before, to institutions. This has meant that the research worker has to conform to the prevailing frames of judgment in order to obtain support from the granting authorities. His application can be considered an offer or tender which, if successful, may lead to a contract. He must then carry out the assignment if he is to reckon with renewal of funding or support for further work.

These research offers are judged by peer review groups according to a scale that allows comparisons of different applications. However, both the peers and the applicants have one common interest: to guarantee the delivery of meaningful information and minimize the risk of fruitless research. This inevitably means that priority is given to research that derives from established premises and where results can be expected.

What we achieve in this way is a rewarding of certain abilities: experimental skill, the ability to pose relevant and intelligent questions, and the capacity to fulfill the task. This type of research is valuable and essential. However, the danger in this form of selection is that it leads easily to an overemphasis on research areas which already have impeccable theoretical structure and even favors those who cater to the structure of the peer review committees themselves. Under such conditions one will not give priority to projects that are difficult to formulate within the predominating conceptual framework. Let us look a little more closely at this aspect. One may divide medical research into strategic or tactical research and curiosity research. Strategic research is motivated by societal or industrial needs or by topical developmental research in areas that are very active and where financial support is likely. Curiosity research is motivated by a desire to know. For the reasons discussed previously, strategic or tactical research is often given priority today. But can we really afford to ignore curiosity research?

The answer must be no. Curiosity is the cradle of creativity, and, when combined with imagination, enthusiasm, critical thought, and stubbornness, can lead to creative discoveries. These may in turn cause a leap in the flow of knowledge that drives development forward with a bound. Such a breakthrough is always followed by a period rich in developmental research. Let us take examples: Crick and Watson's curiosity about the structure of the DNA molecule led to enormous development; Yalow and Berson's intense interest in the antigenicity of insulin started a revolution in biology; Earl Sutherland's investigation of a cyclic nucleotide led to a great breakthrough in cell biology.

What are the conditions for creativity in the scientific world today? Allow us some personal thoughts. The creative process starts as a tension between the still-unformulated idea and the formulated conceptual framework. When the idea is formulated, its validity must still be tested. To accomplish this, excellence is invaluable, even if it is not entirely necessary that the originator of the idea be gifted with this excellence. It is here that the research group which possesses a wide insight into the relevant world of knowledge and possesses high excellence has its real value, as we shall see later.

What are the factors that stimulate creativity in research? Let us approach this problem by examining the conditions under which the discoveries were made for which we awarded Nobel Prizes. Such an analysis assumes that the scientific discoveries that led to Nobel Prizes were based on creative reasoning. What conditions fostered this creativity?

Table 1 shows the hegemony of Germany in Nobel Prizes in the natural sciences during the first 25-year period, a relatively equal division among the three dominating nations between the years 1926 and 1950, and a remarkable dominance for the U.S. during the period 1951–1976. It is remarkable how a country like Great Britain has maintained its standing during the whole period, while Germany has fallen behind considerably. The standing of the nations has remained largely unchanged during the last 15 years since the economic consequences of World War II need no longer be considered a factor.

What clues does this distribution of laureates provide? It is tempting to conclude from these figures that Germany's scientific society has been thinned out since the war, or has to a great degree directed its activities to developmental research. The opposite would then be true of the U.S. One could also conclude that the climate for research and its economic support has worsened in Germany and improved in the U.S. Perhaps factors other than these have contributed to the pronounced change. One widely held opinion is that American research gained from the immigration of established European scientists during the years 1930–1941. There is hardly any ques-

**Table I.** Distribution of the Nobel Prizes for Natural Sciences among the Large Nations

| | Distribution during 25-year periods | | |
|---|---|---|---|
| Country | 1901–1925 | 1926–1950 | 1951–1976 |
| USA | 3 | 14 | 43 |
| Germany | 19 | 13 | 7 |
| Great Britain | 9 | 11 | 19 |
| France | 10 | 4 | 2 |

tion that the eight European scientists who were forced to move to the U.S. after receiving the Nobel Prize had an enormous impact on later advances made in their special fields: the physicists Niels Bohr, Albert Einstein, Enrico Fermi, James Franck, and Victor Hess; the chemist Peter Debye, the pharmacologist Otto Loewi, and the biochemist Otto Meyerhof. Three others, the physicists Wolfgang Pauli and Otto Stern and the biochemist Henrik G. P. Damm, were to receive the prize shortly after immigration. Most of these scientists were Germans. With them came other leading members of the research elite who also made their impact on research in the U.S. by their own contributions and by founding schools. If we examine those eight Nobel laureates who were forced to flee to the States, we find a patrimony of their pupils who became laureates.[1] All this can be regarded as a gift from occupied Europe to America: a gift that comprised not only present and future Nobel laureates, but also the resource for a whole new generation of scientists.

In this way Central Europe lost a tradition in learning, which obviously takes more than a generation to build up again. By contrast England, since her scientific schools remained almost intact throughout the strife, was able to maintain her scientific standard at an elite level and after the war even to increase her share of Nobel laureates. The migration of some of the scientific celebrities of Central Europe to England undoubtedly also contributed. It is also noteworthy that many of the scientific giants who left Central Europe were Jews, as were many of those who would later belong to their research teams. We may assume that an enormous number of Jewish research workers of less renown, and those just beginning their research careers, also had to flee Europe.

In this respect it is interesting that studies by Lipsett and Ladd[2] have shown that the Jewish minority in America is disproportionately represented among collegians and even more so among professors. The Jewish representation among American Nobel laureates is even more dramatic.[1] We may suspect that the high proportion of Jewish laureates in the U.S. is in keeping with the Jewish tradition of seeking higher education and its ready availability in America.

It is almost a paradox that no one from Israel has ever won the Nobel Prize in medicine, physics, or chemistry. The tradition of learning naturally must be found also in Israel. Can one of the reasons be that no research worker of Nobel Prize potential fled to Palestine–Israel during that critical decade and there founded a germinal center for research at the ultra-elite level? Or (more likely) is the reason that in the newly established state of Israel—just as in the reestablished lands of West Germany and Japan after the war—economic and other reasons forced scientists to give priority to more pressing developmental research?

We must draw the conclusion that a tradition in learning, however important, cannot alone be the fertile ground for creative scientific research. Neither can economic resources guarantee creativity. The further important factor that must be present is the mental and social climate that encourages the formation of research centers with sufficient density of knowledge and intellectual skills. The structure of such centers is decisive for their ability to carry on creative research. Not all leading scientists are capable of creating germinal centers for research. To achieve this, a scientist must possess some special qualities, e.g., generosity, the capability to recognize differing talents, a broad-minded attitude toward dissidents and questioning, and a love of teaching. When such ideally composed research groups appear, it is extremely important that they get every possible support in their work. Such an investment involves fewer risks than money delegated to project research.

Of course there will always appear research workers with creative gifts outside the established institutions, like Einstein in his patent office and Békésy in his telephone exchange. Such exceptional talents will always come to the fore.

Naturally there are factors that repress creativity in research. One factor already mentioned is the complexity of our world of knowledge. This gives free and wide scope to workers who are receptive and gifted verbally but lack originality—those who never question. The research that such people accomplish binds economic resources which could be put to better use. We hope that the majority of research groups with their wide common resources of knowledge and inherent analytical capacity can see through less meaningful research problems.

Before World War II research workers' resources of time and

money were very limited while their social image was very respectable. After 1945, when society realized that research substantially contributed to socioeconomic, technical, and medical developments, a period of rapid, almost romantic expansion was launched. However, today for the first time we experience society's doubts that research will be able to achieve peaceful solutions to the problems of mankind. Society finds it difficult to draw the line between research and its exploitation for destructive purposes. The mass media call forth a threatening picture of the consequences of scientific development which in turn can lead to stifling of research and alienation of creativity, e.g., the debates around atomic energy and DNA-hybrid research. This antiintellectualism is an expression of waning confidence, especially among young people, in the future and the possibilities of research to benefit man's estate. We cannot deny that research can be used for any possible purpose, but it should be remembered that science in itself is neither good nor evil. Despite everything, it is the aims, social awareness, human attitude, and principles of the scientist that are good or evil.

## REFERENCES


1. Zuckerman, H., 1980, *Scientific Elite: Nobel Laureates in the United States,* Free Press, New York.
2. Lipsett, S. M., and Ladd, E. C., 1971, Jewish academics in the United States: Their achievements, culture and politics, *American Jewish Yearbook,* pp. 89–128.

# 2

# Basic Biomedical Research
## A Cost–Benefit Analysis

H. HUGH FUDENBERG

## INTRODUCTION

Data analyzed in a previous volume of this series[1] strongly suggested that investment of federal funds in basic biomedical research results not only in fiscal benefits to American taxpayers, but also in improvements in the quality of health care available to them. By analyzing the dollar savings from fundamental research findings in the past which eventually led to improvements in health care (e.g., new, better, and less costly methods for diagnosis, treatment, prophylaxis, and/or eradication of disease), I estimated that taxpayers are currently saving \$20–\$30 yearly for each dollar spent 5–20 years ago on basic biomedical research in just one area, immunology.[2]

To those readers participating in biomedical research, the need for the funding of basic investigations seems self-evident; however, in the face of unprecedented federal budget cuts in almost all nonmilitary areas, the value of "ivory tower" research into fundamental mechanisms underlying biological phenomena is being overlooked both by the taxpaying public and by their elected representatives in Washington. Since this country's largest source of biomedical research funds is the budget of the National Institutes of Health (NIH), a budget which must be approved by Congress, it appears necessary for the scientific community to prove to Congress that federal funding

H. HUGH FUDENBERG • Department of Basic and Clinical Immunology and Microbiology, Medical University of South Carolina, Charleston, South Carolina 29425.

of basic biomedical research results not "merely" in lives saved but also in dollars saved. This chapter reviews three such arguments: first, a "scientific" justification for the support of basic biomedical science, based on the original writings of Drs. Julius Comroe and Robert Dripps[3]; second, an economic argument based on corporate investment patterns in research and development and their effects on long-term growth in productivity; and third, an updated look at the fiscal returns of biomedical research.

Dr. Lewis Thomas, probably the most articulate spokesman for bioscience, has divided biomedical research into two types: one asks *what if* and the other *how to*. *What if* and *how to* are terms which he (and I) believe preferable to the common designations of "basic" versus "applied" (or "targeted") research. For example, some NIH funds allocated for health-care delivery so that newer methods of health care can be made available to a greater percentage of the population are devoted to finding out *how to* use more effectively the knowledge already available. Funding for this type of research must be distinguished from NIH research grants awarded to scientists to support their investigations of areas which in some cases may have no immediate or apparent relevance to the health of American taxpayers. The data presented here are intended to show that investments in the latter type of research—the *what if*—provide even greater returns in the long run.

## SCIENTIFIC BASIS FOR THE SUPPORT OF BIOMEDICAL SCIENCE

In 1966 the Department of Defense published a preliminary report of a study conducted by a team of scientists and engineers, with the aim of analyzing retrospectively how 20 important military weapons came to be developed. Their study, Project Hindsight, was used by the Johnson administration during the late 1960s as an argument for targeted, as opposed to basic, research funding. Among the study's conclusions were the following: (1) the contributions of university research were minimal; (2) scientists contributed most effectively when their effort was mission oriented; and (3) the time between initial discovery and final application was shortest when scientists' work was targeted by their sponsors.[4]

The report of Project Hindsight had a great impact on Congress and on the Office of Management and Budget. Perhaps not surprisingly the same conclusions were generalized to include biomedical research, and a sharp upsurge of contract and commission-initiated (*how to*) research supported by NIH soon followed. Indeed since 1966 there has been a continuing debate as to whether the federal government gets more for those biomedical research dollars used to support clinically oriented research as compared with those used to support biomedical research that is not clinically oriented (i.e., the biological analogue of "more bang for a buck").

In an attempt to introduce a scientific approach to the debate over applied versus basic biomedical research, Comroe and Dripps undertook an analysis of clinical advances in the field of cardiovascular and pulmonary diseases.[3] To ensure that their study was concerned directly with the health of the nation rather than with "esoteric" scientific discoveries, they directed their attention only to those clinical advances (since the early 1940s) that had been directly responsible for diagnosis, prevention, or cure of cardiovascular or pulmonary disease, for stopping its progression, or for prolonging the useful lives of its victims. To avoid their own bias, they asked 40 prominent physicians (engaged in medical practice rather than research) to list those advances they considered to be the most important in terms of the medical care available to their own patients. They classified each selection as either cardiovascular or pulmonary, and a list of the advances selected in each area was sent to 40–50 specialists in that field for the selection of the top 10 advances (Table I). Then, with the help of 140 consultants, they identified the "essential bodies of knowledge" that had provided the scientific basis for each of the 10 advances selected as being among the most important in cardiovascular and pulmonary medicine and surgery in the past 30 years.

Their concept of "essential bodies of knowledge" can be illustrated for the first of the top 10 advances, cardiac surgery. Although general anesthesia was first used in 1846, the first successful operation on an open heart with complete cardiopulmonary bypass apparatus was not performed until 1954, 108 years after the first use of

**Table I.** The Top 10 Clinical Advances in Cardiovascular and Pulmonary Medicine and Surgery in the Last 30 Years

| |
|---|
| Cardiac surgery (including open-heart repair of congenital defects and replacement of diseased valves) |
| Vascular surgery (including repair or bypass of obstruction or other lesions in aorta, coronary, cerebral, renal, and limb arteries) |
| Drug treatment of hypertension |
| Medical treatment of coronary insufficiency (myocardial ischemia) |
| Cardiac resuscitation, defibrillation, "cardioversion," and pacing in patients with cardiac arrest, slow hearts, or serious arrhythmias |
| Oral diuretics (in treatment of patients with congestive heart failure or hypertension) |
| Intensive cardiovascular and respiratory care units (including those for postoperative care, coronary care, respiratory failure, and disorders of newborns) |
| Chemotherapy and antibiotics (including prevention of acute rheumatic fever and treatment of tuberculosis, pneumonias, and cardiovascular syphilis) |
| New diagnostic methods (for earlier and more accurate diagnosis of diseases of the cardiovascular and pulmonary-respiratory systems) |
| Prevention of poliomyelitis (especially of respiratory paralysis due to polio) |

ether anesthesia. The reason for the delay was that a number of other bodies of knowledge, in addition to anesthesia, were essential before surgeons could successfully and predictably repair cardiac defects. For example, a precise preoperative diagnosis is required for the patient; but this required selective angiocardiography, which required the earlier discovery of cardiac catheterization, which required the still earlier discovery of x rays. An artifical heart–lung apparatus (pump–oxygenator) is also needed to take over the function of the patient's heart and lungs during the operation; this required not only a pump designed so that it would not damage the blood and, for oxygenators, a basic understanding of the exchange of $O_2$ and $CO_2$ between gas and blood, but also the discovery and purification of a potent, nontoxic anticoagulant (e.g., heparin). For successful open-heart surgery, Comroe and Dripps listed a total of 25 essential bodies of knowledge. For the top 10 advances they identified 137 essential bodies of knowledge.

The next step in their study was the identification of "key articles" that were particularly important to the development of one or more of the 137 essential bodies of knowledge. About 4000 published articles were examined, of which about 2500 specific scientific reports were selected. These were arranged chronologically in 137 tables from which, with the advice of their consultants, Comroe and Dripps selected more than 500 essential or key articles for careful study. This admittedly tedious and time-consuming approach produced more data which serve to emphasize that scientific advance requires far more work than that reported by the "discoverer," or even by those who wrote the key articles needed for his discovery. The chronological tables compiled by Comroe and Dripps show scientific events and the dates of the publications reporting them (key articles); for example, 46 separate events in the development of electrocardiography were identified.[3] Thus these tables provide specific evidence for policy-makers that scientists earlier and later than the discoverer have always been essential to each discovery and to its full development, and that good, original research of the *what if* type provides the basis for later advances in practical medical knowledge and improvements in the quality of life.

The questions ultimately asked in the study were "Was the key research clinically oriented?" and "Was the key research basic or not?" First, each key article was classified as clinically oriented if the author mentioned in the article, even briefly, an interest in or possible application of his findings to the diagnosis, treatment, or prevention of a clinical disorder, or to explaining the basic mechanisms operative in the disease state; key articles in which the authors did not state or suggest any bearing of their research on human disease were classified as not clinically oriented, even though the work may later have been applied to some aspect of a clinical disorder in humans. Using this criterion, Comroe and Dripps calculated for each of the top 10 advances in cardiovascular and pulmonary medicine (Table I) the percentage of key articles leading to the advance that were not clinically oriented (Table II). Although the percentage varies depending on the advance selected, it is clear that in every case a significant amount of the essential work was done by scientists who expressed (at least in writing) no interest in a clinical problem at the time their work was reported. These data strongly suggest that planning for future clinical advances *must* include generous support for innovative and imaginative *what if* research that has no immediate relevance to a clinical problem at the time the work is funded.

**Table II.** Goal of Authors of 529 Key Articles That Later Were Judged to Be Essential for a Clinical Advance

| Clinical advance | Clinically oriented | Not clinically oriented | Total | Percent of total not clinically oriented |
|---|---|---|---|---|
| Cardiac surgery | 53 | 35 | 88 | 39.8 |
| Vascular surgery | 40 | 8 | 48 | 16.7 |
| Hypertension | 35 | 44 | 79 | 55.7 |
| Coronary insufficiency | 44 | 21 | 65 | 32.3 |
| Cardiac resuscitation | 24 | 16 | 40 | 40.0 |
| Oral diuretics | 19 | 24 | 43 | 55.8 |
| Intensive care[a] | — | — | — | — |
| Antibiotics | 40 | 13 | 53 | 24.5 |
| New diagnostic methods | 41 | 53 | 94 | 56.4 |
| Poliomyelitis | 16 | 3 | 19 | 15.8 |
| Total | 312 | 217 | 529 | 41.0 |

[a]A key article was assigned to only one advance even though it may have been essential to more than one. Because practically every key article in intensive care was also essential to other advances, these articles were assigned elsewhere (for example, to cardiac or vascular surgery, coronary insufficiency, resuscitation, or antibiotics).

Finally, in order to answer the question of whether key research was basic or not, Comroe and Dripps chose as their definition of basic research any investigation in which an attempt was made to determine the mechanisms responsible for the effects observed, as opposed to simply observing, describing, and measuring effects. The 529 key articles identified earlier in their study were analyzed to determine how each investigator carried out his research. Each article was placed into one or more of six categories:

1. Basic research unrelated to the solution of a clinical problem.
2. Basic research related to the solution of a clinical problem.
3. Studies not concerned with basic biological, chemical, or physical mechanisms.
4. Review and critical analysis of published work and synthesis of new concepts (without new experimental data).
5. Developmental work or engineering to create, improve, or perfect apparatus or a technique for research use.

6. Developmental work or engineering to create, improve, or perfect apparatus or a technique for use in diagnosis or care of patients.

The results of classifying the 529 key articles above into these six categories are shown in Table III. Of 567 entries (some key articles fit into more than one of the categories), 209 were in category 1 and 141 in category 2; in other words, by the criteria of the study 61.7% of the total entries were reports of basic research. According to this analysis (Table III) basic research was responsible for nearly three times as many key articles as other types of research, and indeed nearly twice

**Table III.** Types of Research Reported in 529 Key Articles

| | Basic | | | | Development | | |
|---|---|---|---|---|---|---|---|
| Type | Not clinically oriented | Clinically oriented | Not basic | Review and synthesis | Research | Clinical | Total |
| Cardiac surgery | 34 | 23 | 19 | 0 | 3 | 11 | 90 |
| Vascular surgery | 9 | 7 | 14 | 3 | 0 | 21 | 54 |
| Hypertension | 42 | 16 | 21 | 2 | 0 | 0 | 81 |
| Coronary insufficiency | 21 | 20 | 22 | 1 | 1 | 3 | 68 |
| Cardiac resuscitation | 16 | 11 | 9 | 0 | 0 | 6 | 42 |
| Oral diuretics | 23 | 13 | 6 | 1 | 0 | 0 | 43 |
| Intensive care[a] | — | — | — | — | — | — | — |
| Antibiotics | 12 | 18 | 21 | 1 | 0 | 2 | 54 |
| New diagnostic methods | 49 | 21 | 5 | 2 | 17 | 22 | 116 |
| Poliomyelitis | 3 | 12 | 3 | 0 | 1 | 0 | 19 |
| Total | 209 | 141 | 120 | 10 | 22 | 65 | 567[b] |
| Percent of total | 36.8 | 24.9 | 21.2 | 1.8 | 3.9 | | |

[a]Because practically every key article in intensive care was also essential to other advances, these articles were assigned elsewhere (for example, to cardiac or vascular surgery, coronary insufficiency, resuscitation, or antibiotics).

[b]The total number of entries in the six categories (567) exceeds the total in Table II (529) by 38 entries. This is because some key articles fit into more than one category here, particularly when articles reporting development of new apparatus also reported research using it; no article in Table II was classified more than once.

as many as the combined total for both nonbasic research and developmental work. Thus basic research, as defined by Comroe and Dripps, is shown by their "scientific" analysis to pay off in terms of key discoveries almost twice as handsomely as other types of research and development combined.[3]

The pioneering work of Comroe and Dripps in analyzing how clinical advances result from biomedical research has been widely recognized among scientists as the best scientific argument currently available to support the contention that federal investment in *what if* studies should continue at least at its present level.[5,6] However, the budgets proposed by the Reagan administration for the coming years (see below) will mean that both the National Science Foundation (NSF) and NIH will be forced to reduce their support of research and development of all kinds, because the budgets proposed for those agencies do not cover the shrinking value of the dollar caused by inflation. I believe that more studies of the kind carried out by Comroe and Dripps, undertaken perhaps with the sponsorship of various scientific societies, would help deflect further cuts. I also suggest that the scientific community should become more vocal in pointing out to the taxpaying public and to their elected representatives in Washington that federal funds invested in basic biomedical research produce long-term payoffs in the form of improvements in the quality of health care available in the United States and, as argued below, in dollar savings.

## THE ECONOMIC IMPORTANCE OF BASIC RESEARCH

In 1978 at the time of my earlier article[1] the nation's total annual health bill was estimated to be in the neighborhood of $160 billion; yet the total proposed NIH budget (of which only about one-third was devoted to basic research) was only $2.8 billion, or less than 2% of the total U.S. health costs for the previous year (for calendar 1981, $267 billion). I suggested that those figures indicated "fiscal irresponsibility" on the part of the federal government, and that any corporate managers running a $160-billion-a-year enterprise would devote at least 5% of their total budget to research. An analysis of the patterns of research investment by major U.S. corporations provides, at least in my opinion, strong support for those statements. Recently President François Mitterrand of France announced that he is counting on an impressive commitment of resources to science and technology to help propel France out of the economic doldrums and to give his country's high-technology industry a competitive future in work markets. The major aims of the government's plan for science and technology are embodied in a new research law that commits the government to a 5-year program of increases in the science budget including a pledge to raise R&D expenditures from less than 2% of the gross national product last year to 2.5% of GNP by 1985. In addition to these specific targets the law sets out a number of more general objectives aimed at creating a new national framework for research by promoting improved interactions among industry, government laboratories, and universities. Strong efforts will also be made to make the French more conscious of the importance of science and technology in their lives and to involve them in decisions on science and technology.[7]

This economic argument for basic research has support in an impressive number of past studies. For example, Dr. Edwin Mansfield conducted a survey of changes in productivity in both agricultural and manufacturing industries during the 1960s. His findings showed that, without exception, an industry's or firm's rate of productivity increase was directly related to the amount spent on research and development.[8] Similar conclusions have been reached in studies by Griliches,[9] who analyzed data from almost 900 manufacturing firms, and Terleckyj,[10] whose data for entire industries indicated that an industry's rate of productitivity increase is directly related to both the amount of its own research and development and the amount carried out by industries that supply and support it (since research by one industry often leads to improved equipment or processes for another, related industry). For instance, predictions of high profits at Texas Instruments exist for its 64K RAM, the well-publicized computer memory that stores some 64,000 items of data on a silicon chip the size of a fingernail.

More recently, Mansfield's studies[11] have provided evidence that the *composition* of an industry's or firm's research and development expenditures, as well as their magnitude, influences the rate of productivity increase. He found that there is a direct relationship between the amount of *basic research* carried out by an industry or firm and its rate of productivity increase when its expenditures on applied research on development are held constant.

Table IV shows the 1979 figures for total sales, investments in research and development, and profits for major U.S. corporations in the appliance, steel, automotive, and information-processing industries. The first three industries are those currently experiencing the greatest financial difficulties; in general U.S. corporations in these areas have spent the dollar equivalent of only about 1% of their sales on research. This is perhaps because their chief executive officers

**Table IV.** Investments in Research and Development by U.S. Industries[a]

| Industry | Company | 1979 sales ($ billions) | Research and development expense (% of sales) | Profits (%) |
|---|---|---|---|---|
| Appliances | Hoover | 0.754 | 1.0 | 18.2 |
| | Singer | 2.598 | 1.3 | −35.2 |
| | Sunbeam | 1.323 | 1.4 | 37.2 |
| | Whirlpool | 2.261 | 1.5 | 30.1 |
| Steel | Allegheny Ludlum | 1.550 | 0.9 | 20.4 |
| | Bethlehem Steel | 7.137 | 0.6 | 14.9 |
| | Republic Steel | 3.987 | 0.5 | 15.6 |
| | U.S. Steel | 12.929 | 0.4 | −14.8 |
| Automotive | American Motors | 3.117 | 1.9 | 87.8 |
| | Chrysler | 12.002 | 3.0 | −32.7 |
| | Ford Motor | 43.514 | 4.0 | 147.1 |
| | General Motors | 66.311 | 2.9 | 67.4 |
| Information processing | Amdahl | 0.300 | 13.0 | 227.6 |
| | Data General | 0.507 | 10.0 | 101.6 |
| | General Automation | 0.116 | 6.9 | 691.0 |
| | IBM | 22.863 | 5.9 | 45.2 |
| | Beckman Instruments | 0.427 | 7.8 | 113.1 |
| | Teradyne | 0.122 | 10.4 | 159.4 |
| | Advanced Micro Devices | 0.148 | 7.3 | 99.4 |
| | American Micro Systems | 0.109 | 8.1 | 204.3 |
| | Intel | 0.663 | 10.1 | 85.8 |
| | Intersil | 0.139 | 8.0 | 108.3 |
| | National Semiconductor | 0.720 | 9.4 | 198.3 |

[a] *Source: Business Week*, July 7, 1980, p. 49.

(usually accountants or lawyers by training) derive their compensation in large part in the form of annual bonuses or stock options based on the profits for the preceding year and thus they are not interested in what will happen in their industries 15 years down the road. This is in marked contrast to Japanese companies in the same industries, where the managers (usually Ph.D.s), chemists, or engineers are employed for life and are evaluated on the basis of long-term (10–15 years) results; thus they have a great interest in innovation and perfection of new technologies for use 10–15 years hence (for other comparisons of Japanese to U.S. industry see below).

The major pharmaceutical firms in the U.S. have done relatively well; they spend about 5%–10% of their total sales dollars on research. Even more striking, as shown in Table IV, corporations in the most advanced technological industry, information processing (microprocessors, silicon chips, computer hardware), which have enjoyed the greatest economic success in recent years, spend an average of 10%–15% of their total sales dollars on research. The chief executive officers of the latter companies are almost invariably scientists (rather than accountants or lawyers), and their rewards are determined by long-term increases in the value of their companies' stocks, rather than by immediate compensation.

Thus it appears that fundamental research in industry produces innovative findings which, when later applied by that industry, result in large dollar benefits (i.e., corporate profits). The same holds true, I believe, for the nation's health-care industry. Unfortunately the percentage of the nation's health-care budget devoted to biomedical research falls far short of the figure that would be expected, on the basis of the above data, to provide long-term economic stability in the health-care industry. In 1980 the federal government spent $70.9 billion on health care, and total spending for health care in the U.S. during that calendar year was estimated at $247.2 billion, a record 9.4% of the gross national product.[12] The annual U.S. health bill for 1982 is now estimated at approximately $267 billion (out of a total estimated gross national product of approximately $3 trillion). The total NIH budget for this year is estimated at only about $3.5 billion, and nearly half of that figure is for health care (technology transfer) and/or training, not for basic research (Table V). That is, we are still

**Table V.** Federal Investments in Research and Development at the National Science Foundation (NSF) and National Institutes of Health (NIH)[a,b]

| FY | Total R&D | Basic research | Applied research | Development |
|---|---|---|---|---|
| NSF | | | | |
| 1979 | 807,925 | 733,255 | 66,770 | 7,900 |
| 1980 | 881,792 | 815,246 | 58,764 | 8,105 |
| 1981 (est.) | 937,404 | 873,640 | 59,764 | 4,000 |
| 1982 (est.)[c] | 999,900 | 945,444 | 54,456 | — |
| NIH | | | | |
| 1979 | 2,953,133 | 1,463,703 | 1,066,408 | 423,022 |
| 1980 | 3,181,830 | 1,642,341 | 1,145,129 | 394,360 |
| 1981 (est.) | 3,328,345 | 1,745,645 | 1,188,645 | 394,055 |
| 1982 (est.)[c] | 3,570,821 | 1,883,650 | 1,270,400 | 416,771 |

[a] *Source:* National Science Foundation, 1981, *Federal Funds for Research and Development, Fiscal Years 1980, 1981, and 1982: Detailed Statistical Tables,* Volume 30, NSF, Washington, D.C.
[b] Amounts shown are in thousands of dollars.
[c] Figures for 1982 are "lower limit" amounts, based on the budget for FY 1982 submitted by the president to Congress in March of 1981. These figures are subject to congressional approval and/or revision.

spending less than 1.5% of health-care costs on fundamental biomedical research, an amount proportionally far below those for the industries that are now setting standards for economic competition both in the U.S. and internationally.

As suggested above, Japanese computers in most industrial areas and U.S. high-technology microelectronics companies have fared relatively well compared to U.S. automotive, appliance, and steel companies. In recent years even American microelectronics firms have been experiencing great difficulty because of far greater research efforts and funding by the Japanese. In 1977, according to the electrical industries association of Japan, Japanese "electronic firms ploughed nearly 30% of their sales revenue back into R&D and new plant. In 1982, they will invest almost 40%."[13] In the autumn of 1980 64K RAM chips were introduced in Japan; the cost was approximately $100 apiece. The cost was reduced in 1981 to less than $10, and chips now sell for $5 each.[14] Japan, supplier of over 70% of the 8–10 million or so 64K RAM chips sold worldwide, is a year ahead of California.[15] Indeed the California-based high-technology company

Hewlett-Packard is preparing to begin making silicon chips based on designs and technology supplied by Japan's Hitachi.[16] In automation Japan led in robot production by a gigantic figure (14,200 compared to the U.S. with 4000).[17] The Japanese government announced early this year a 3-year budget of $45 million to begin research of "fifth-generation" computers which possess artificial intellegence (AI): "This computer project will have special import because it most likely will be equipped with rudimentary intelligence and may use light instead of electricity to function. It thus will assume an increasing share of the research burden."[18] Actual development of the initial fifth-generation precedents will cost approximately $450 million over a 10-year period. However, the potential fiscal benefits of such research are enormous. AI applications will soon lead the computer field in both importance and financial profits: "Japanese computer makers hope to establish themselves as the leaders in the commercialization of AI and thus leapfrog their US competitors." On the U.S. side AI research was begun by Schlumberger, Ltd., the oil services giant, over 5 years ago, employing dozens of scientists in three separate laboratories, but its estimated AI budget is only $5 million annually. However, Schlumberger's chairman, Jean Riboud, has categorized AI research as critical technology, stating, "Artifical intelligence is as important to our future as the surge in oil exploration."[19]

The budget of the Japanese Ministry of International Trade and Industry (MITI) will include research for the development of a lightning-fast supercomputer (an $85-million project) and for a so-called optical computer, using optical fibers (costing $80 million). The majority of MITI's funds will be awarded in 100% grants to industry. However, many Japanese firms have already begun research into supercomputers; indeed, Fujitsu, now Japan's leading computer manufacturer, will announce the production of its own supercomputer later this year. These supercomputers are about a tenth the size of general-purpose mainframe computers and will use radically different integrated circuits capable of incredibly more rapid computing than today's most sophisticated microchips.

"A fifth-generation" machine[20] of the 1990s will have the ability to "infer." Instead of merely manipulating data or words, it will process knowledge—writing its own programs, deciding for itself which steps to take to solve a given problem, automatically retrieving information from a library, extracting rules of its own from facts, and then applying its new findings to solve problems of which humans may be unaware.

Japan's MITI also initiated a 10-year biotechnology research program of $110 million. Indeed an estimated 20 billion-dollar Japanese companies (e.g., Mitsubishi Chemical Industries) are now involved in biotechnology research, compared to the small maiden companies in the U.S. 5 years ago (e.g., Genex, Molecular Genetics, Inc.). A belated entrant into biotechnology, Japan was slow in developing techniques for gene-splicing and other technology invaluable to scientists in modifying the genetic structure of living organisms. However, through concerted efforts Japan is rapidly catching up to U.S. biotechnology. Japanese researchers and businessmen have been visiting U.S. technology firms and are approaching these U.S. companies asking to license U.S. technology. Simultaneously Japanese companies are expanding their in-house efforts by increasing research staff and the purchase of new equipment. For example, U.S. firms such as Vega Biotechnology, Inc., expect to sell as many of their automatic gene synthesizer machines to the Japanese as they will sell in the U.S.[21]

Again, it appears that the scientific community has failed to impress upon federal policy-makers the lessons learned from the above studies. If nothing is done to change the current pattern of federal support for basic biomedical research, we should not be surprised in the near future to find the nation's health-care industry facing the same sorts of economic problems that now besiege our steel, appliance, and automotive industries.

## FISCAL RETURNS OF BIOMEDICAL RESEARCH: AN UPDATE

During the first year of the Reagan administration's new fiscal policy of cuts in federal spending in nondefense areas, demands on the federal budget and competition for the funds available have increased to new levels. Under these conditions it is particularly important for scientists to provide convincing arguments in support of federal investment in fundamental biomedical research programs. As I suggested in the past,[2] justification of biomedical research in terms of improvements in quality of life or even in terms of lives saved no longer seems sufficient to convince the public and their elected representatives in Congress that such expenditures deserve high priority. Rather, it appears that expenditures for biomedical research must be justified in view of taxpayer dollars saved if such research is to be supported by the general public. In previous articles[2,22,23] I cited examples of research in areas supported by grants from the National Institute of Allergy and Infectious Diseases (NIAID) that have resulted in fiscal savings. At that time only about 5% of the total NIH budget was devoted to this institute (now 7.5%), which supports research in virologic diseases, bacteriologic diseases, parasitic diseases, fungal diseases, allergy, and autoimmune diseases, plus almost all of the basic research in both immunology and recombinant DNA, currently the two most rapidly expanding of the various biologic disciplines. Examples of fiscal savings from past research in these areas include the following:

1. Polio has been eradicated, with dollar savings of approximately $2 billion yearly (Table VI) in 1960 dollars. Studies of polio began at Rockefeller University in 1910. By 1938 investigators had found that monkeys who had never had polio could be immunized by injections of serum from monkeys who had recovered from polio, mixed with polio virus. In the late 1940s the National Foundation for Infantile Paralysis began a crash program to cure polio. Little was accomplished toward a cure, however, until a method of virus propagation *in vitro* in monkey kidney in tissue culture was discovered by investigators doing basic research into virus propagation. The techniques developed by J. F. Enders and his colleagues Thomas Weller and Frederick Robbins, who later received a Nobel Prize for their work, saved valuable time and eliminated the need for live monkeys in polio research. Subsequently, after the Salk vaccine had been developed and introduced, basic research into the immunology of the gastrointestinal tract demonstrated that an orally administered vaccine

**Table VI.** Estimated Losses Avoided in the United States through the Prevention of Paralytic Poliomyelitis during the Period 1955–1961, and Estimated Cost of Avoidance

| Estimate | Millions of dollars |
|---|---|
| Losses avoided | |
| Medical care costs | 326.8 |
| Gross lifetime income lost | 6389.7 |
| Total | 6716.5 |
| Cost of avoidance | |
| Vaccine purchase | 129.8 |
| Physician's vaccination fees | 468.6 |
| Vaccine administration | 13.3 |
| Government-funded research and field trials | 41.3 |
| Total | 653.0 |
| Net gain | 6063.5 |

produces greater resistance than one administered intramuscularly or subcutaneously and leads to better protection against polio. Studies by a number of research groups showed that "secretory immunoglobulins"—antibodies formed in the intestinal tract—play an important role in preventing reinfection by viruses that enter and multiply in the respiratory and enteric tracts. Ultimately these *what if* research findings enabled development of a successful oral polio vaccine, and by the early 1960s polio was defeated.

2. A rubella (measles) vaccine has been developed, resulting in prevention of the congenital deformations that occur in children of pregnant mothers who develop the disease. Such children often require lifelong institutionalization; the fiscal savings that resulted from eliminating the need for their care are estimated at about $180 million yearly.

3. Basic immunologic studies designed to learn the mechanism of antibody feedback control have resulted in data which were applied to the eradication of Rh hemolytic diseases in newborns (by the administration of small amounts of anti-Rh antibody to Rh-negative mothers) and thus have eliminated this

serious illness, which in the past resulted in significant fetal and neonatal mortality and furthermore caused mental impairment of sufficient degree to necessitate lifelong institutionalization for a significant percentage of its victims. Estimated savings from the eradication of this disease are about $60 million yearly in the United States, and at least ten times that throughout the world.

4. By immunologic means, the antigen associated with transfusion hepatitis (TH) has been identified, so that symptom-free donors who are silent carriers of the virus can now be screened and their blood not used for transfusion purposes. Such protection is eliminating sufficient cases of chronic hepatitis so that dollar savings now total approximately $100 million yearly.

   More recently the Food and Drug Administration has announced approval of the first vaccine against TH to be licensed for use in the United States.[23] Some 200,000–300,000 Americans each year are infected with the virus, of whom 50,000–60,000 fall ill with jaundice, fever, and nausea, and 10,000 require hospitalization; approximately 100–200 Americans die of acute hepatitis each year. The Centers for Disease Control estimate that the virus costs the nation $750 million annually in medical bills and time lost from work, and even those who are only mildly infected run the risk of becoming chronic carriers of hepatitis B and developing cirrhosis or liver cancer later in life. Development of this vaccine, which became available for general use in mid-1982, began in 1968. It is the first completely new viral vaccine in 10 years and the first vaccine ever licensed in the United States that is made directly from human blood. (It also promises to protect against post-hepatitis liver cancer, a very common disease in the Orient.)

5. In the field of transplantation immunology, as a result of fundamental research on genetically determined antigens on membranes of immunologic cells, and as a result of research on the effects of antitumor drugs on the immune response in mice, subsequent scientific developments have delineated the

genetic differences responsible for rejection of kidney transplants, using a technique called the "mixed leukocyte culture," which originally was thought to represent the test tube equivalent of graft–host reactions. Through close matching of donors and recipients by these newly developed immunologic techniques, and through administration of immunosuppressive drugs in correct amounts, transplantation is replacing dialysis. Estimated cost savings are approximately $100 million yearly.

6. Since February 1978 a pneumococcal vaccine containing capsular material from 14 types of pneumococci has been marketed in the United States. It has been estimated that as many as 75% of all cases of pneumococcal pneumonia are caused by these 14 types. Pneumonia is the fifth leading cause of death in the United States, and estimates of the extent of pneumococcal pneumonia range from 5000 to as high as 66,000 deaths each year; on the basis of the lower estimate of 5000 deaths annually, the Office of Technology Assessment has estimated that treatment of pneumococcal pneumonia cost at least $135 million in 1978.[24] Presumably an effective vaccination program will lead to savings of approximately that amount in future years.

Even more important in terms of current funding decisions by the president and Congress are the potential savings in future years from the advances now being made in immunology. Other new vaccines are under development and should be in wide use in the relatively near future. For example, scientists trying to develop vaccines against the diverse bacteria that cause meningitis are making significant headway. A new tetravalent vaccine is available against meningococci, one group of meningitis-causing bacteria, which will make it possible to prevent 40%–45% of the cases now caused by meningococci. A combined vaccine against meningococcus group B is also being tested now, and efficacy results will be available this year; this group of bacteria strikes mostly children under 2 years old and is responsible for 50% of meningococcal disease. Meningococci were responsible for 3454 U.S. cases of meningitis reported in 1981.[25] Almost twice as many cases are caused each year by another bacterium, *Haemophilus influenzae*, and a vaccine developed at the University of Rochester has been shown in recent tests to Finland to protect children over 18 months old; however, since babies are the prime target for this organism, further research is needed.

The first effective vaccine against hepatitis B, described above, resulted from collaboration by investigators at the NIAID and other institutes at NIH, in addition to investigators in industry and university research groups. In 1982 and subsequent years NIAID funds will support the production of a second generation of vaccines, drawing on recombinant DNA technology and peptide synthesis to circumvent the present reliance on antigens harvested from the blood of carriers.

Recent advances in microbiology and immunology have led to technical innovations such as recombinant DNA and hybridoma cell fusion. Such advances provide new opportunities to accelerate the development of new vaccines in the future. Fundamental studies on etiologic agents, identification of immunogenic antigens, and the immunogenetics of antibody responses to vaccines should allow the application of these technical advances in new strategies for the development of vaccines and eradication of other infectious diseases, with similar fiscal returns.

In addition, hybridoma technology has a vast number of other potential applications. The hybridoma technique, developed by Cesar Milstein and George Kohler in 1975, is probably among the most important inventions of the decade. This technique—based on 20 years of basic immunologic research on antibody production, cell culture, and cell fusion, and antibody production *in vitro*—provides a way of obtaining a line or "clone" of identical cells which manufacture a single or "monoclonal" antibody directed against a chosen target. Besides their use in research, monoclonal antibodies have already proved to be excellent tools for diagnosis of various diseases, and their applications to therapy (e.g., therapy of human neoplasia) are now being explored.[26] It has been suggested that the worldwide market for monoclonal antibodies could be half a billion dollars by 1987.

The cells of the immune system are divided into many different sets and subsets, some enhancing and others suppressing the immune response. The fine structure and function of molecules that serve as regulators of these complex interactions are being defined by current immunologic research, using monoclonal antibodies, lymphocyte products known as lymphokines, and cloned immunologically reactive cells. One possible application of the knowledge gained from such research may be to restore imbalances in the interrelationships among the lymphocytes in various immunologic diseases. For example, multiple sclerosis (MS) has been shown to involve an imbalance in subsets of T lymphocytes[27] (perhaps either the cause or a result of a viral disease or some neuroimmunologic disorder); the cost of this disease alone has been estimated at $500,000 over the lifetime of each male patient, and approximately $2 billion annually in the United States, based on a 1972 National MS Society survey.[28]

Equally important for long-range savings is current research on the genetic mechanisms that determine the immune responses to pathogenic invaders, cancer cells, transplanted organs, and environmental substances that cause allergies. The genes that control these immune responses are termed the *major histocompatibility complex* (MHC). Recognition of the clinical association between the MHC and many diseases—including juvenile-onset diabetes, juvenile rheumatoid arthritis, chronic active hepatitis, and myasthenia gravis—now offers exciting possibilities for understanding the mechanisms by which disease susceptibility is inherited. Knowledge of these mechanisms could lead to the development of strategies for supplementing our natural resistance against these and other diseases and for increasing the effectiveness of vaccination programs, as well as to more effective methods for retention of lifesaving organ and tissue transplants.

According to my admittedly nonscientific estimate, based on the above and other examples of past and potential future fiscal savings, the dollar benefit–cost ratio for funds spent by NIAID on immunologic research may be at least 100 : 1. In my opinion the strategy of the Office of Management and Budget (OMB) in attempting to balance the federal budget by reducing federal investment in biomedical research must be regarded as "penny wise, pound foolish." The most cost-saving method of health-care delivery is eradication of disease; the most likely way to accomplish this is through increased fundamental research, which at the time of funding may have no apparent relationship to any known disease, but makes possible the creation of knowledge and technology for future application.

Probably the most dramatic example in recent years of the potential fiscal benefits derived from the application of basic research findings is the use of microorganisms whose DNA has been altered by insertion of foreign genetic material which instructs the organism to produce new and different proteins. Recombinant DNA technology is already being used for production of proteins of potential significance in medicine, agriculture, animal husbandry, and, in the near future, energy technology (for example, to clean up oil spills); for the production of marketable substances; or for methodology derived from fundamental research in microbial genetics.

Consider somatostatin, a hormone involved in the production control of insulin and growth hormones. In 1971 when somatostatin was discovered, it took nearly half a million sheep brains to produce just 5 mg of it; now the same amount can be grown in a 2-gallon broth of genetically modified bacteria. Many other examples are now familiar to most readers. And yet 20 years ago bacterial genetics and molecular biology were considered among the most "ivory tower" of all the disciplines in basic biomedical research, and were generally believed to have no clinical or other "worthwhile" applications.

The origins of molecular biology lie in early medical research on the bacteria that cause pneumococcal pneumonia. Dr. Oswald Avery began his research on pneumonia at the Rockefeller Institute for Medical Research in 1913. Thirty years later he and his colleagues published a paper identifying deoxyribonucleic acid (DNA) as the substance that induces genetic "transformation" of pneumococcal types.[29] His work was followed by that of Watson and Crick, who determined the structure of the DNA molecule in 1953.[30] In the past 30 years molecular biology has consisted almost entirely of exploring the genetic machinery of one species of bacteria, *Escherichia coli*, and of the viruses (bacteriophages) that attack *E. coli*. Although it might seem impractical, perhaps even irrational, to study the viruses of bacteria when viral diseases in humans remain to be conquered, it turned out that bacteriophages were useful probes for examining bacterial genes. It was discovered that certain enzymes (restriction endonucleases) can be used to locate and cut out specific genes, insert them into small DNA circles (plasmids), and seal the ends of the circle to form a new plasmid containing the inserted gene (which can be derived from any species, including man). This modified plasmid can be inserted into another *E. coli* cell, which will reproduce the newly inserted gene in each of its progeny. Billions of *E. coli* are produced, each carrying the modified plasmid and making the product coded by the inserted gene.

Recombinant DNA technology has already been used to produce human insulin (for the treatment of diabetes), somatostatin (as described above), and interferon, and the potential for other applications in the immediate future is enormous. Indeed, according to economists who have surveyed the potential, the estimated world market for the products of recombinant DNA—whether for medical use, low-cost production of chemicals (such as antifreeze), creation of heat-resistant and frost-resistant corn and other crops, production of gasohol, clean-up of sewage, or a host of other uses—will be more than $200 billion yearly within 10 years (see Chapter 3).

Table VII shows a summary of the research projects supported by NIH in four areas related to the development of recombinant DNA technology for the 5-year period ending in 1978 (after that date a considerable part of the research in those four areas consisted of projects using the technology rather than research or development projects on the "science base" for the technology). According to this estimate, generated by Dr. John Nutter at NIAID using a central NIH computer-based system (CRISP), the total NIH investment in the development of the technology was less than $100 million during this 5-year period. The total amount invested by federal agencies, pharmaceutical and other corporations, the American Cancer Society and other private funding agencies, and various venture-capital firms has been estimated at approximately ten times that amount ($1 billion) over the past 5 years, much of which was devoted to development of applications (*how to*) rather than fundamental investigations (*what if*). Clearly, compared with an estimated future market for recombinant

**Table VII.** NIH Science Base for Recombinant DNA Technology[a]

| FY | Lambda phage | Plasmids | *E. coli* K-12 | Endonuclease | Total |
|---|---|---|---|---|---|
| 1974 | $ 2,617 | $ 1,574 | $1,201 | $ 1,637 | $ 7,029 |
| | (53) | (35) | (11) | (26) | (125) |
| 1975 | 3,213 | 2,788 | 643 | 3,433 | $10,077 |
| | (52) | (51) | (16) | (62) | (181) |
| 1976 | 3,693 | 5,600 | 496 | 5,934 | 15,723 |
| | (56) | (71) | (10) | (87) | (224) |
| 1977 | 5,515 | 7,474 | 1,561 | 10,549 | 24,735 |
| | (60) | (82) | (20) | (145) | (307) |
| 1978 | 6,537 | 9,316 | 1,408 | 16,728 | 33,989 |
| | (79) | (119) | (21) | (221) | (440) |
| Total | $21,211 | $26,752 | $5,309 | $38,281 | |
| | (300) | (358) | (78) | (541) | |
| Grand total | | | | | $91,553 |
| | | | | | (1,277) |

[a]In thousands of dollars; number of projects is given in parentheses. Estimates generated using a central NIH computer-based system (CRISP) which uses various terms to describe research projects. The four terms used first appeared as a group in FY 1974, which corresponds nicely with the emergence of the technology. After FY 1978 the search would tend to identify projects using the technology rather than research or development projects on the science base for the technology. Data provided by John E. Nutter, Ph.D., OPPE, NIAID.

DNA products of $200 billion yearly, the investment in basic biomedical research which led to their development is now yielding impressive fiscal benefits.

Another promising area of research in molecular homology of viruses has resulted in the development of new antiviral compounds. For example, one new compound, Ara-A, has been used successfully to treat a number of life-threatening viral infections, including herpes encephalitis, which cannot be prevented with available vaccines. Another antiviral drug, acyclovir, administered to immunosuppressed patients who are to receive bone marrow transplants, apparently can prevent the development of herpes simplex virus infections, a common complication after such therapy. Clinical improvement was also observed in patients with herpes genitalis, and, although acyclovir does not eradicate the latent infections, it appears to provide prophylaxis against reactivated infections and to decrease dissemination of the virus to sexual contacts. Research on this and other antiviral compounds (such as interferon) may eventually provide relatively inexpensive pharmaceuticals that will be effective against a wide variety of viruses, including the common cold (rhinovirus).

Recently an ingenious assay for the antenatal diagnosis of sickle cell anemia has been devised. Utilizing a restriction endonuclease (which cleaves the DNA coding for the abnormal portions of the globin chain) and DNA hybridization techniques, the test requires as little as 10–20 ml of amniotic fluid and may be done on uncultured fluid cells. The availability of this simple, accurate, and safe assay has led to a crescendo of tests for this disease (15 in 6 months) as compared to 6 tests from 1974 to 1978. The benefits of genetic counseling based on the knowledge gained from such tests in alleviating the cost of treating sickle cell anemia are obvious.[31]

In addition, as more DNA sequences associated with particular genetic diseases are cloned in bacteria, it is expected that similar types of hybridization analyses will allow definition of the patterns of inheritance more clearly and allow prenatal diagnosis of more diseases. These techniques will be particularly suitable for diseases that cannot already be diagnosed through biochemical analysis of amniotic fluid or accessible fetal blood cells. Also, this technique has detected small amounts of hepatitis B viral DNA in liver-biopsy specimens from patients with recurrent hepatitis and hepatoma. This approach is likely to aid in the diagnosis of many other viral diseases.[32]

## CONCLUSION

Estimates published by the OMB show that funds for basic research across all government agencies would grow by some 12% over the comparable FY 1981 level if the FY 1983 budget request of the OMB is accepted; however, further analysis indicates that these figures represent no real growth, since percentage increases in the consumer price index for the same period amount to *at least* 21%. The recommended FY 1983 level for NIH is only 3% above current funding and actually below the FY 1982 level originally approved by appropriations committees of Congress—$70 million below the Senate figure and $86 million below the House figure (Table VIII). Despite the fact that biomedical research is often cited as an administration

**Table VIII.** Total NIH Budget Appropriations by Institute and Research Division[a]

| Institute | FY 1980 | FY 1981 | FY 1982[b] | FY 1983[c] |
|---|---|---|---|---|
| Cancer | $ 998.0 | $ 990.6 | $ 943.0 | $ 955.4 |
| Heart, lung, blood | 527.1 | 550.4 | 559.6 | 577.1 |
| Dental | 67.6 | 71.3 | 72.0 | 74.5 |
| Arthritis | 340.1 | 369.9 | 368.2 | 379.4 |
| Neurology | 241.4 | 252.9 | 265.9 | 274.5 |
| Allergy | 214.7 | 232.4 | 235.9 | 246.0 |
| General medical sciences | 312.3 | 334.1 | 335.5 | 345.6 |
| Child health | 208.3 | 220.9 | 226.4 | 233.6 |
| Eye | 109.6 | 118.1 | 127.4 | 131.6 |
| Environmental health | 83.6 | 93.6 | 154.3 | 157.4 |
| Aging | 69.7 | 75.8 | 81.9 | 84.6 |
| Research resources | 169.1 | 175.8 | 184.2 | 191.0 |
| Fogarty Center | 8.7 | 9.1 | 9.2 | 10.1 |
| Subtotal | 3,350.2 | 3,494.9 | 3,566.1 | 3,661.8 |
| National Library of Medicine | 43.9 | 44.7 | 44.4 | 46.0 |
| Office of director | 20.8 | 21.7 | 22.6 | 24.3 |
| Buildings and facilities | 13.9 | 11.7 | 9.9 | 17.5 |
| Total | 3,428.8 | 3,569.4 | 3,643.0 | 3,750.0[d] |

[a]Dollar amounts in millions.
[b]December 1981 Continuing Resolution, based on the administration's reductions, which involved a 4%–6% cut plus a $7 million reduction for the NIH in administrative costs, exclusive of transfers.
[c]Amounts approved by congress, released January 1983.
[d]NIH reported that there will be a drop of 20% in the number of new and competing renewal grants funded under the Reagan administration budget request for FY 1983 compared with FY 1981, from 5109 down to 4100. In the same two-year period there will be a drop of 17% in the number of research trainees, from 10,695 down to 8900, although there has been a growing consensus within government and the scientific community on the need to stabilize the number of trainees at 10,000 and the number of new and competing renewal grants at 5000.

priority, the total FY 1983 recommendation for NIH of $3.7 billion is only $101 million above the current level, an increase of only 3%, well below the predicted 6%–8% increases in the rate of inflation for FY 1982–1983.

When the administration's budget proposal for FY 1983 was first presented, the OMB announced that total federal expenditures on research and development were scheduled to rise from $38.8 billion in FY 1982 to 43.0 billion in the coming year—an increase of 10.8%. However, closer analysis shows that military programs would account for the entire increase, whereas nondefense (including biomedical) research and development would get level funding; in real terms, taking into account the effects of inflation, support for civilian science and technology would decrease. Indeed while military research and development has experienced real growth (after inflation) of 22.2% between FY 1980 and FY 1982, federal support for civilian research and development has declined by 16.1%.[33]

The OMB is requesting cutbacks in two areas of NIH programs: competing research grants and training awards. Ironically these are the areas which most directly affect the nation's commitment to basic biomedical research. According to the OMB proposal the number of competing research grants would be reduced by more than 10%, from 4700 last year to 4100 in FY 1983. This seems to make meaningless the target number of 5000 grants agreed to only 3 years ago by the Department of Health and Human Services in order to stabilize long-range planning of research.[34] The OMB budget sets aside $492 million

**Table IX.** Key Members of 1983 Congressional Committees and Subcommittees That Legislate Biomedical Research Budgets

| |
|---|
| *U.S. Senate* |
| Committee on Appropriations—Mark O. Hatfield (R–Ore.), Chairman |
| Agriculture and Related Agencies—Thad Cochran (R–Miss.) |
| Energy and Water Development—Mark O. Hatfield |
| Labor, Health and Human Services, and Education—Orrin G. Hatch (R–Ut.) |
| Committee on the Budget—Pete V. Domenici (R–N.M.), Chairman |
| Committee on Environment and Public Works—Robert T. Stafford (R–Vt.), Chairman |
| Environmental Pollution—John H. Chafee (R–R.I.) |
| Nuclear Regulation—Alan K. Simpson (R–Wyo.) |
| Toxic Substances and Environmental Oversight—Slade Gorton (R–Wash.) |
| Water Resources—James Abdnor (R–S.D.) |
| Committee on Labor and Human Resources—Orrin G. Hatch (R–Utah), Chairman |
| Aging, Family, and Human Services—Jeremiah Denton (R–Ala.) |
| Alcoholism and Drug Abuse—Gordon J. Humphrey (R–N.H.) |
| Education, Arts, and Humanities—Robert T. Stafford (R–Vt.) |
| Employment and Productivity—Dan Quayle (R–Ind.) |
| Handicapped—Lowell P. Weicker (R–Conn.) |
| Special Committee on Aging—John Heinz (R–Pa.), Chairman |

*(Continued)*

**Table IX** *(Continued)*

*U.S. House of Representatives*

Committee on Appropriations—Jamie L. Whitten (D–Miss.), Chairman
- Agriculture—Jamie L. Whitten
- Defense—Joseph P. Addabbo (D–N.Y.)
- Energy and Water Development—Thomas Bevill (D–Ala.)
- HUD-Independent Agencies—Edward P. Boland (D–Mass.)
- Interior—Sidney R. Yates (D–Ill.)
- Labor, Health, and Human Services—William H. Natcher (D–Ky.)
- Transportation—William Lehman (D–Fla.)

Committee on the Budget—James R. Jones (D–Okla.), Chairman

Committee on Education and Labor—Carl D. Perkins (D–Ky.), Chairman
- Elementary, Secondary, and Vocational Education—Carl D. Perkins
- Employment Opportunities—Augustus F. Hawkins (D–Calif.)
- Health and Safety—Joseph M. Gaydos (D–Pa.)
- Human Resources—Ike F. Andrews (D–N.C.)
- Postsecondary Education—Paul Simon (D–Ill.)
- Select Education—Austin J. Murphy (D–Pa.)

Committee on Science and Technology—Don Fuqua (D–Fla.), Chairman
- Energy Development and Applications—Don Fuqua
- Energy Research and Production—Marilyn L. Bouquard (D–Tenn.)
- Investigations and Oversight—Albert Gore, Jr. (D.–Tenn.)
- Natural Resources, Agricutural Research, and Environment—James H. Scheuer (D–N.Y.)
- Science, Research, and Technology—George E. Brown, Jr. (D–Calif.)
- Space Science and Applications—Ronnie G. Flippo (D–Ala.)
- Transportation, Aviation, and Materials—Dan R. Glickman (D–Kan.)

Select Committee on Aging—Claude D. Pepper (D–Fla.), Chairman

for competing grants, a drop of $35 million, or 7%, from last year. (In order to get funds for 500–600 of those 4100 grants, NIH will reduce indirect cost reimbursement by 10% of the current rate.) The OMB also asked that the number of research trainee awards be decreased from 9700 to 8900 next year; expenditures for these awards would total $152 million, or a decline of 3%. This cut will diminish the numbers of the next generation of biomedical scientists below the "critical level" (12,500) each year recommended in a 4-year study by a National Research Council "blue ribbon committee" on health science manpower needs.

The NIH budget currently proposed by the OMB reflects neither sound scientific planning nor sound fiscal policy, as shown by the benefit–cost analyses presented above. I believe that the biomedical research community can lobby successfully for congressional approval of a budget that will be more realistic in meeting current and future biomedical research needs and thus in reducing the future costs of health-care delivery in this country. It is hoped that the arguments outlined here will be brought to the attention of those members of Congress whose votes will decide the fate of the nation's biomedical research (Table IX), either through scientists' action committees (e.g., The Science and Technology Political Action Committee, P.O. Box 351, Rockville, MD 20850) or through the individual efforts of concerned scientists.

## ACKNOWLEDGMENT

I am grateful to Dr. H. D. Whitten for editorial assistance and to H. Rhett Coombs Halverson for invaluable aid in collecting economic data and references.

## REFERENCES

1. Fudenberg, H. H., and Melnick, V. L. (eds.), 1978, *Biomedical Scientists and Public Policy*, Plenum Press, New York.
2. Fudenberg, H. H., 1978, Informing the public: Fiscal returns of biomedical research, *in: Biomedical Scientists and Public Policy* (H. H. Fudenberg and V. L. Melnick, eds.), Plenum Press, New York, p. 35.
3. Comroe, J. H., Jr., and Dripps, R. D., 1978, Scientific basis for the support of biomedical science, *in: Biomedical Scientists and Public Policy* (H. H. Fudenberg and V. L. Melnick, eds.), Plenum Press, New York, p. 15.
4. Sherwin, C. W., and Isenson, R. S., 1966, *First Interim Report on Project Hindsight*, Office of Director of Defense Research and Engineering, Washington, D.C., June 30, revised October 13).
5. Garfield, E., 1979, How can we prove the value of basic research?, *Current Contents* **22**(40):5.
6. Garfield, E., The economic impact of research and development, *Current Contents* **26**(51):5.
7. France readies new research law, 1982, *Science* 216:712.
8. Mansfield, E., 1972, Contribution of research and development to economic growth in the United States, *Science* **175**:477.

9. Griliches, Z., 1980,? *in: New Developments in Productivity Measurement and Analysis* (J. Kendrick and B. Vaccara, eds.), National Bureau of Economic Research, Chicago, p. 419.
10. Terleckyj, N., 1974, *Effects of R&D on the Productivity Growth of Industries: An Exploratory Study,* National Planning Association, Washington, D.C.
11. Mansfield, E., 1980, Research and development, productivity, and inflation, *Science* **209:**1091.
12. Federal cost cutting pushes up deficits of nonprofit hospitals, 1982, *Wall Street Journal,* February 9, p. 1.
13. Silicon Valley has a big chip about Japan, 1982, *The Economist,* March 20, p. 20.
14. Japan's ominous chip victory, 1981, *Fortune,* December 14, 1981.
15. Chipmaking expertise, 1982, *Fortune,* April 5.
16. Alexander, T., 1982, Teaching computers the art of reason, *Fortune,* May 17, pp. 82–92.
17. Introducing the one-megabit microchip, 1982, *The Economist,* May 1, p. 93.
18. Gregory, G., *et al.* Japan's "Third Revolution," 1982, *World Press Review,* March, p. 23.
19. Artificial intelligence: The second computer age begins, 1981, *Business Week,* March 8.
20. Will Japan leapfrog America on superfast computers?, 1982, *The Economist,* March, 6, p. 95.
21. Government boost for biotechnology industry, 1982, *Business Week,* December 14.
22. Fudenberg, H. H., 1972, The dollar benefits of biomedical research: A cost analysis, *J. Lab. Clin. Med.* **79:**353.
23. Fudenberg, H. H., 1974, Fiscal returns of biomedical research, *Am. Rev. Respir. Dis.* **109:**411.
24. Willems, J. S., Sanders, C. R., Riddiough, M. A., and Bell, J. C., 1980, Cost effectiveness of vaccination against pneumococcal pneumonia, *N. Engl. J. Med.* **303:**553.
25. Vaccine makers gain on meningitis, 1982, *Med. World News,* February 15, p. 30.
26. Hybridomas: The making of a revolution, 1982, *Science,* February 26, p. 1073.
27. Goust, J. M., Hoffman, P. M., Pryjma, J., Hogan, E. L., and Fudenberg, H. H., 1980, Defective immunoregulation in multiple sclerosis, *Ann. Neurol.* **8:**526.
28. Waksman, B. H., 1980, Multiple sclerosis, the mystery disease, *Infectious Diseases Newsletter,* **10**(12):4.
29. Avery, O. T., MacLeod, C. M., and McCarty, M., 1944, Studies on the chemical nature of the substance inducing transformation of pneumococcal types: Induction of transformation by deoxyribonucleic acid fraction isolated from pneumococcus type III, *J. Clin. Invest.* **179:**137.
30. Watson, J. D., and Crick, F. H. C., 1953, A structure for deoxribose nucleic acid, *Nature* (*New Biol.*) **171:**737.
31. Chang, J. C., and Kan, Y. W., 1982, A sensitive new prenatal test for sickle cell anemia, *N. Engl. J. Med.* **307:**30–32.
32. Chang, J. C., Golbus, M. J., and Kan, Y. W., 1982, Antenatal diagnosis of sickle cell anemia by sensitive DNA assay, *The Lancet* **I:**1463.
33. *Science News,* 1982, January 16, p. 37.
34. *FASEB Newsletter,* 1982, **15**(4):1.

# 3

# Molecular Biology

## Application to Prenatal Diagnosis of Genetic Disorders of Hemoglobin

YUET WAI KAN

To some people, scientists included, the advances in molecular biology that have been hailed from time to time in the popular and scientific press have yet to live up to the promises made for them. But practical benefits there are, and three recent articles[1–3] illustrate the progress being made in another direction—the antenatal diagnosis of thalassemia and sickle cell disease, serious disorders that affect millions of people in some parts of the world. While the goal of treatment is still elusive, increasingly ambitious prevention programs are being undertaken and the social impact is already becoming apparent.

In times of economic austerity, funding for biomedical research tends to favor "clinical" over "basic" research. Often research funds are targeted toward a specific disease or group of diseases, as if somehow the channeling of resources will produce immediate results. This policy is short-sighted since many discoveries initiated by investigators solely in the pursuit of basic knowledge have had far-reaching "practical" applications. A striking recent example of the contribution basic research can make to clinical medicine is prenatal diagnosis of the inherited hemoglobin abnormalities. This chapter reviews the development of methods for prenatal diagnosis of sickle cell anemia and

YUET WAI KAN • Howard Hughes Medical Institute Laboratory and Department of Medicine, University of California at San Francisco, San Francisco, California 94143.
The text of this chapter was first published in *The Lancet*, November 21, 1981, and it is reprinted here with revisions.

the thalassemias, and also discusses the impact that prenatal tests have had on families at risk for these diseases. Savings in health-care costs are also analyzed. Prenatal diagnosis of the hemoglobinopathies is still in its infancy, and although significant short-term benefits have already been realized, the long-term advantages have yet to be seen.

## THE DISEASES IN QUESTION

The primary function of hemoglobin is to transport oxygen from the lungs to the tissues. The major hemoglobin molecule in the adult, hemoglobin A, is comprised of two pairs of globin chains which combine to form the 4-chain $\alpha_2\beta_2$. If abnormal or insufficient globin chains are produced, a hemoglobin disorder results. Of the many inherited hemoglobinopathies defined, sickle cell anemia and thalassemia are the most common. Both of these diseases cause major health-care problems and have pronounced social and economic effects in many parts of the world.

Sickle cell anemia is found primarily in people of African origin, but also occurs in the Mediterranean area and parts of Asia.[4,5] The inherited condition, which is caused by a mutation that results in production of an abnormal β-globin chain, is characterized by lifelong hemolytic anemia. The abnormal hemoglobin polymerizes when deoxygenated; the red blood cells assume an abnormal "sickle" shape and are prematurely destroyed. These abnormal red cells often block blood vessels, causing acute pain and organ damage. The common causes of infant mortality are infection, severe anemia, and thrombosis of the blood vessels of the central nervous system. In older patients cardiac and renal failure are frequent causes of death. The clinical manifestations of sickle cell anemia are extremely variable; some patients have recurrent debilitating blood vessel clotting episodes, while others manifest a much milder disease. It is not possible to predict the severity of the disease.

The common feature of the thalassemias is inadequate production of a normal component of hemoglobin.[6] Thalassemia arises from a variety of genetic mutations including gene deletions, abnormal processing of messenger RNA, and nonsense mutations. The most common thalassemia syndromes are the α- and β-thalassemias, in which synthesis of the α- and β-globin chains of adult hemoglobin are respectively impaired. α-Thalassemia is seen in many countries, but the severe homozygous form of the disease occurs predominantly in Southeast Asia. An affected fetus dies *in utero* during the second half of pregnancy, or immediately after birth. β-Thalassemia is prevalent throughout the world and constitutes a major health problem in the Mediterranean area, the Middle East, and Southeast Asia. Patients affected by homozygous β-thalassemia (also known as Cooley's anemia) appear healthy at birth. Within a few months to a few years they become extremely anemic and require frequent blood transfusions to survive. Until recently β-thalassemia homozygotes usually died at a young age from anemia or infection, although some lived into their twenties when they succumbed to cardiac failure. However, new methods of management have increased the life expectancy and enhanced the quality of life for β-thalassemia homozygotes. These methods include more vigorous transfusion regimens designed to maintain the hemoglobin at nearly normal levels, and daily subcutaneous continuous infusions of desferioxamine, a drug that binds excess iron and excretes it in the urine. The treatment programs prevent organ damage from the anemia and iron overload, and enable many patients to experience normal physical development. However, the transfusion requirement, daily injections, and recurrent hospitalizations impose a great burden, both emotional and financial, on the patient and his or her family. Since this type of therapy has only been available within the past 10 years, it is not yet possible to assess its long-term effects on the survival of these patients.

## GENETIC SCREENING AND COUNSELING

Sickle cell anemia and β-thalassemia are inherited according to Mendelian laws. If both parents are "carriers" of these diseases (each has one abnormal gene), there is a 1 : 4 chance that their offspring will inherit both abnormal genes and suffer from the homozygous disease. Carriers are asymptomatic, but can be detected by blood tests. Theoretically if carriers are detected at an early age they can be counseled about their condition and advised against choosing a spouse who is also a carrier. In practice this type of counseling has not been effective and in fact has several serious drawbacks.[7] First, it necessitates testing every individual before or at reproductive age. Setting up such a massive screening of a large segment of a population is extremely complicated, both in terms of the expense and the logistics involved, and can only be contemplated in a small, relatively homogeneous population group. Second, a close monitoring of one such screening and counseling program revealed that the mating patterns of the carriers remained basically unchanged despite the program. Third, the information presented at a genetic counseling session is often forgotten, or in some cases even misinterpreted. The sickle cell screening programs instituted in the early 1970s illustrate what can happen when programs are not adequately supported by counseling and public education. Many carriers were left with the mistaken impression that they actually had a disease, some faced job discrimination, and others were charged high insurance premiums.

Previous experience with other genetic disorders such as Tay–Sachs disease indicates that screening and counseling are most effective when accompanied by a prenatal diagnosis program.[8] The screening can be targeted to individuals of childbearing age in high-risk ethnic groups. If both partners are found to be carriers, they can be offered the option of prenatal diagnosis.

## METHODS OF PRENATAL DIAGNOSIS

Amniocentesis was introduced as a method for diagnosis of hereditary disorders in the fetus in the late 1960s.[9] Over 50 abnormalities can now be detected prenatally with this procedure. It has become a safe and established technique, performed in medical centers throughout the world. At approximately the 15th week of gestation a sample of amniotic fluid is taken from the amniotic cavity, and, depending on the disease in question, either the amniotic fluid or the cells (which are both fetal in origin) are analyzed. For example, Down's syndrome, a common abnormality affecting children born to mothers over 35, is detected by analyzing the chromosome composition of the amniotic fluid cells; some neurotube defects such as spina bifida are diagnosed by determining the α-fetoprotein content of the fluid; and many inherited enzyme defects can be tested by analyzing the enzyme content of the amniotic fluid cells.

The conventional method of diagnosing sickle cell anemia and the thalassemias is to analyze the type of hemoglobin present in the red blood cells. Since the fetal cells in amniotic fluid are composed of skin and other epithelial cells which do not synthesize hemoglobin, amniocentesis does not provide a usable sample for hemoglobin analysis. In the early 1970s methods were developed to obtain and analyze small samples of fetal blood.[10] The placenta is first localized with ultrasound and a fiber-optic instrument (fetoscope) is inserted into the uterus to allow direct visualization of a blood vessel on the placenta or in the umbilical cord. A blood sample is aspirated with a small-gauge needle contained in the instrument. The size of the red cells is monitored by a particle size analyzer, which differentiates between the large fetal cells and the smaller adult red cells. Both sickle cell anemia and β-thalassemia can be diagnosed by determining the type (for sickle cell disease) or the quantity (for β-thalassemia) of β-globin chains synthesized by the red blood cells.

Fetoscopy is still the most commonly used method for prenatal diagnosis of β-thalassemia. A major disadvantage of the procedure is that it requires specialized training, available to few obstetricians. Another more serious drawback is that, unlike amniocentesis, fetal blood sampling carries a definite risk of fetal mortality. When the procedure is first instituted in a medical center the fetal mortality rate often exceeds 10%, although it usually decreases to 3%–5% after the obstetrician has acquired more experience. Furthermore, the procedure has an extremely limited availability. In the U.S., for example, only two medical centers—in New Haven and San Francisco—are equipped to perform the procedure. Patients must travel to these centers, often at great expense or over long distances, to undergo the test.

The first step in the chain of events that lead to a clinical application for DNA analysis techniques was the discovery of restriction endonucleases in the early 1970s.[11] These are enzymes that cut (cleave) DNA at a particular nucleotide sequence. Over 100 restriction enzymes have been characterized, each cleaving the DNA at a specific sequence. The next step occurred in 1975 when Southern devised a technique for analyzing the genome of higher organisms using restriction enzymes.[12] Finally, in the late 1970s his method was adapted to study globin genes in humans.[13,14] Human DNA is first digested with a restriction enzyme which generates anywhere from a few hundred thousand to several million DNA fragments of different lengths. These fragments are separated according to size by electrophoresis on a gel support medium. They are then transferred to and immobilized on special filters. A radioactive probe specific for the globin gene under investigation is synthesized *in vitro* and used as a hybridization probe for the DNA fragments. Only the fragments that contain globin genes will hybridize with the probe. The filters are exposed to x-ray film, and the globin gene-containing fragments are identified. Each enzyme yields a characteristic and reproducible restriction pattern.

Several approaches have been used to locate the defect in the DNA in thalassemia and sickle cell anemia.[15–17] If the mutation responsible for a defect affects a long stretch of nucleotides, it may alter the restriction patterns of several different enzymes, and abnormal patterns will be generated when the DNA is digested with these enzymes. Sometimes the mutation is much smaller, involving only a few nucleotides or even a single nucleotide. If the mutation happens to occur precisely at a DNA sequence that is the recognition site for a particular restriction enzyme, it can abolish a cleavage site for that enzyme or create a new one. Thus digesting the DNA with such an enzyme will produce distinctly different patterns in normal and mutant DNA.

Often the exact nature or location of a mutation is unknown, or even if the mutation has been defined, it may not involve a restriction enzyme recognition site. In these instances the genetic defect can be detected indirectly by linkage analysis. This is based on the principle that if another readily detectable mutation is closely linked on a chromosome to the gene of interest, it can be used as a marker to indicate whether or not the gene under investigation is normal. The mutation used for linkage analysis lies in a region of DNA near the globin gene, either in another globin gene or in a segment of the DNA that does not encode for any known protein.

Sickle cell anemia was originally detected by indirect linkage analysis.[15] In a region of DNA near the β-globin gene (where the sickle mutation is located) lies a neutral mutation which happens to abolish a normal recognition site for the enzyme Hpa I. Individuals with the sickle mutation who originate from West Africa have lost this Hpa I site. Thus when their DNA is digested with Hpa I, the fragment that contains the β-globin gene differs in length from normal DNA. In these individuals this difference in fragment length serves as a marker for prenatal diagnosis since it indicates the presence or absence of the sickle gene. The test is extremely sensitive and can be performed on amniotic fluid DNA prepared directly from uncultured amniotic fluid cells, thus eliminating the necessity of fetoscopy. However, it cannot be applied to all pregnancies at risk for sickle cell disease because not all sickle genes are associated with the neutral Hpa I mutation. Parents must be screened to determine whether they have the marker. In the American black population the test is useful in about 70% of the pregnancies at risk.

Recently the enzyme Mst II was used to detect directly the sickle cell mutation in amniotic fluid DNA.[1,2,16] In the region coding for amino acids 5–7 in the normal β-globin DNA, the sequence CCTGAGG is recognized by the enzyme Mst II. Cleavage of human β-globin DNA with Mst II generates multiple DNA fragments, including two encompassing this region—one 200 and the other 1150 base pairs in length. Sickle β-globin contains the sequence CCTGTGG, which is not recognized by Mst II. Therefore cleavage of sickle β-globin with Mst II generates only one fragment, 1350 base pairs in length. These differing cleavage patterns can distinguish normal, sickle-trait, and homozygous sickle DNA in cell samples obtained by routine fluid aspiration. Thus this new Mst II strategy allows direct examination of small amounts of amniotic fluid. How widely applicable to other genetic diseases will this approach be? Several thalassemic syndromes can be diagnosed because the mutations causing disease alter restriction-endonuclease cleavage sites.[17] Thalassemia resulting from complete gene deletion can also be detected with an analogous strategy.[18]

Most α-thalassemias are due to α-globin gene deletions.[18] The lethal homozygous form of the disorder is caused by deletion of all the α-globin genes. This condition is therefore quite easy to detect; digesting amniotic fluid DNA with one of several restriction enzymes will show that the fragments that contain the α-globin genes are absent. The diagnosis can be made at mid-trimester, thus eliminating 20 weeks from a pregnancy doomed to produce a nonviable fetus. Termination of these abnormal pregnancies prevents maternal toxemia and postpartum hemorrhage, two complications that often occur with these pregnancies.

Since β-thalassemia is rarely caused by gene deletion, direct DNA analysis has been used successfully in several instances.[19] In Sardinia and China a DNA polymorphism that affects a Bam HI site close to the human β-globin gene can be used to exclude thalassemia in approximately 50% of Sardinian pregnancies at risk, and in about 40% of the Chinese pregnancies. Other polymorphic restriction sites situated close to the β-globin gene have also been used for prenatal diagnosis. These polymorphic sites are recognized by the enzymes Hin dIII, Ava I, and Hinc II. By analyzing for the presence or absence of these sites in families at risk for β-thalassemia, it is possible to identify the chromosomes that carry the β-thalassemia gene and determine if the fetus is affected by the homozygous disease.[20]

Currently the majority of pregnancies at risk for sickle cell anemia are diagnosed by DNA analysis of amniotic fluid, as are *all* cases of α-thalassemia. Fetal blood analysis has been the predominant method for diagnosis of β-thalassemia, but as an increasing number of polymorphic restriction sites are discovered, more cases may be diagnosed by amniocentesis.

## IMPACTS OF PRENATAL DIAGNOSIS OF SICKLE CELL ANEMIA AND β-THALASSEMIA

The impacts or prenatal diagnosis on these two disorders will be discussed separately, since far more data are available on β-thalassemia. Most of the experience has emerged from England, Italy, Greece, and the U.S. Worldwide, over 2000 cases of prenatal diagnosis for β-thalassemia have been performed and close to 500 pregnancies which would have resulted in the birth of a homozygous child have been terminated.[10] Let us examine the economic benefits derived from prenatal diagnosis of this disorder. Taking into account the cost of hospitalization, blood transfusions, and the expensive iron-chelating drug desferioxamine, the annual cost of caring for a patient with homozygous β-thalassemia is about $10,000. (The actual cost per patient depends on the number of hospitalizations for treatment of complications and the cost of health care in each country.) The total annual cost to care for 400 homozygotes would therefore be $4 million. Assuming that each patient has a 20- to 40-year lifespan (and this will probably increase with the improved treatment available) the total cost could reach $160 million.

Compare these figures with the cost of prenatal diagnosis. The cost of fetal blood analysis, including the physician, equipment, and laboratory analysis, is $500–$1000. Since one out of four pregnancies turns out to be affected, four tests are needed to identify one homozygote. Thus it costs $2000–$4000 to identify one homozygote and $800,000–$1.6 million to detect 400 homozygotes. The differences in the cost–benefit ratio of a lifetime-maintenance program compared to a prenatal diagnosis program are obviously enormous.

More important than the monetary savings of prenatal diagnosis are the benefits derived by parents at risk. In the past couples often found out that they were carriers only *after* giving birth to an affected child. Their usual response was to stop having more children because of the 1 : 4 risk of producing another affected child. Prenatal diagnosis allows these couples to plan further pregnancies since they can choose to continue a healthy pregnancy (which occurs three times out of four) or terminate an affected one. Many parents who previously avoided having children can now plan families. In Greece and Italy, countries with a high incidence of β-thalassemia, the reproductive patterns of parents at risk are changing, and more at-risk couples are having children than before the test was available. This positive aspect of prenatal diagnosis cannot be overemphasized. A common misconception among some physicians and laypersons is that the inevitable end result of prenatal diagnosis is an increased abortion rate. This short-sighted view is not substantiated by the results of the prenatal diagnosis programs that have been initiated for the hemoglobinopathies and other genetic disorders. These programs have all shown that more, not fewer, live births occur in at-risk families. Furthermore, if the fetus is normal, the emotional burden of the mother who has had afflicted infants previously is largely relieved.

Currently the impact of prenatal diagnosis on sickle cell anemia is less dramatic. The diagnostic procedure has not been as widely accepted for this disorder as for the β-thalassemia. Several possible explanations for this phenomenon have been suggested. First, the black community has not utilized prenatal diagnosis for genetic disorders to the same degree as other racial groups. The reasons for this are unclear, but appear to be related to the lack of adequate dissemination of information to this population.[21] Second, many of the developing countries in Africa, where sickle cell anemia is extremely common, do not yet have the resources to implement prenatal diagnosis programs. Whereas amniocentesis is a relatively straightforward procedure, the DNA analysis techniques are complex and require a sophisticated knowledge of molecular biology. A simpler test procedure needs to be developed. Third, the lack of acceptance may be related to the safety of the techniques previously used for prenatal diagnosis. The fetal risk associated with fetal blood sampling often causes couples at risk to avoid the test. The availability of a safe method using amniocentesis may favorably influence the acceptance rate. In fact the number of requests for prenatal diagnosis for sickle cell anemia seems to have increased during the 5 years since the amniotic fluid method was introduced. Finally, attitudes regarding aborting a fetus with sickle cell anemia may not be the same as those toward aborting a thalassemia homozygote, since the clinical course of homozygous sickle cell disease is more variable than β-thalassemia, which is predictably severe. Some parents may therefore not want to undergo the test, or if they do, may not want to abort an affected fetus. More time will be required to evaluate the long-term effects of prenatal diagnosis of sickle cell anemia in light of increased public awareness and the availability of a safe procedure.

## CONCLUSION

The history of prenatal diagnosis of the hemoglobinopathies illustrates the close interdependence that exists between basic research and medicine—neither is undertaken solely for its own sake, even if at the time of discovery a finding does not immediately suggest a useful purpose. For example, the technique for chromatographic separation of hemoglobins was devised in the 1960s primarily to study the structure of hemoglobin. The biosynthetic methods using incorporation of radioactive amino acids and separation of the globin chains, which were the cornerstones of the fetal blood test, were originally developed to characterize abnormal hemoglobins and thalassemia. Of course most of these developments would not have been feasible without the advent of new and sophisticated instruments such as the fetoscope, ultrasonograph, and particle size analyzer. These discoveries were the culmination of years of research in various seemingly unrelated disciplines.

We are only beginning to reap the benefits of the DNA analysis techniques. Recombinant DNA technology has enormous potential both for diagnosis and treatment of many genetic disorders. While this chapter has focused on two hemoglobinopathies, other genetic disorders can be diagnosed by the methods described herein. Many other polymorphic DNA markers are being mapped to different regions of the human chromosome. If these markers can be linked to specific disorders, such as cystic fibrosis or Huntington's chorea, they may extend our knowledge of these diseases, as well as provide a means for prenatal diagnosis. Recombinant DNA technology may also provide a tool for treating genetic disorders. Individual genes have already been isolated from the complex human genome. If these genes can be introduced into a defective cell and induced to function, many hereditary diseases could actually be treated. This approach to gene therapy may become a reality within this decade.

## REFERENCES


1. Chang, J. C., and Kan, Y. W., 1982, A sensitive new prenatal test for sickle cell anemia, *N. Engl. J. Med.* **307:**30–32.
2. Orkin, S. H., Little, P. F. R., Kazazian, H. H., Jr., and Boehm, C. D., 1982, Improved detection of the sickle mutation by DNA analysis and its application to prenatal diagnosis. *N. Engl. J. Med.* **307:**32–36.
3. Wilson, J. T., Milner, P. F., Summer, M. E., Nallaseth, F. S., Fadel, H. E., Reindollar, R. H., McDonough, P. G., and Wilson, L. B., 1982, Use of restriction endonucleases for mapping the allele for $\beta^S$-globin, *Proc. Natl. Acad. Sci. USA* **79:**3628–3631.
4. Seargeant, G. R., 1979, *The Clinical Features of Sickle Cell Disease* (A. G. Bearn, D. A. K. Black, and H. H. Hiatt, eds.), North-Holland Publishing Company, Amsterdam.
5. Platt, O., and Nathan, D. G., 1981, Sickle cell disease, *in: Hematology of Infancy and Childhood,* Volume 1 (D. G. Nathan and F. A. Oski, eds.), W. B. Saunders Company, Philadelphia, pp. 687–725.

6. Weatherall, D. J., and Clegg, J. B., 1981, *The Thalassaemia Syndromes,* 3rd ed., Blackwell Scientific Publications, Oxford.
7. Stamatoyannopoulos, G., 1974, Problems of screening and counseling in hemoglobinopathies, *in: Birth Defects* (A. G. Motulsky and W. Lenz, eds.), Excerpta Medica, Amsterdam, pp. 268–276.
8. Kaback, M. M., Zeiger, R. S., Reynolds, L. W., and Sonneborn, M., 1974, Tay-Sachs disease: A model for the control of recessive genetic disorders, *in: Birth Defects* (A. G. Motulsky and W. Lenz, eds.), Excerpta Medica, Amsterdam, pp. 248–262.
9. Milunsky, A., 1979, *Genetic Disorders and the Fetus: Diagnosis, Prevention, and Treatment,* Plenum Press, New York.
10. Alter, B. P., 1979, Prenatal diagnosis of hemoglobinopathies and other hemotologic disease, *J. Pediatr* **95:**501–503.
11. Nathans, D., and Smith, H. O., 1975, Restriction endonucleases in the analysis and restructuring of DNA molecules, *Annu. Rev. Biochem.* **44:**273–293.
12. Southern, E. M., 1975, Detection of specific sequences among DNA fragments separated by gel electrophoresis, *J. Mol. Biol.* **98:**503–517.
13. Mears, J. G., Ramirez, F., Liebowitz, D., Nakamura, F., Bloom, A., Konotey-Ahulu, F., and Bank, A., 1978, Changes in restricted human cellular DNA fragments containing globin gene sequences in thalassemia and related disorders, *Proc. Natl. Acad. Sci. USA* **75:**1222–1226.
14. Orkin, S. H., Alter, B. P., Altay, C., Mahoney, M. J., Lazarus, H., Hobbins, J. C., and Nathan, D. G., 1978, Application of endonuclease mapping to the analysis and prenatal diagnosis of thalassemia caused by globin-gene deletion, *N. Engl. J. Med.* **299:**166–172.
15. Kan, Y. W., and Dozy, A. M., 1978, Polymorphism of DNA sequence adjacent to human β-globin structural gene: Relationship to sickle mutation, *Proc. Natl. Acad. Sci. USA* **75:**5631–5635.
16. Wilson, J. T., Milner, P. F., Summer, M. E., Nallaseth, F. S., Fadel, H. E., Reindollar, R. H., McDonough, P. G., and Wilson, L. B., 1982, Use of restriction endonucleases for mapping the allele for $\beta^S$-globin, *Proc. Natl. Acad. Sci. USA* **79:**3628–3631.
17. Baird, M., Driscoll, C., Schreiner, H., Sciarratta, G. V., Sansone, G., Niazi, G., Ramirez, F., and Bank, A., 1981, A nucleotide change at a splice junction in the human β-globin gene is associated with $\beta^0$-thalassemia, *Proc. Natl. Acad. Sci. USA* **78:**4218–4221.
18. Dozy, A. M., Forman, E. N., Abuelo, D. N., Barsel-Bowers, G., Mohoney, M. J., Forget, B. G., and Kan, Y. W., 1979, Prenatal diagnosis of homozygous α thalassemia, *J. Am. Med. Assoc.* **241:**1610–1612.
19. Kan, Y. W., Lee, K. Y., Furbetta, M., Angius, A., and Cao, A., 1980, Polymorphism of DNA sequence in the β-globin gene region: Application to prenatal diagnosis of $\beta^0$ thalassemia in Sardinia, *N. Engl. J. Med.* **302:**185–188.
20. Antonarakis, S. W., Boehm, C. D., Giardinia, P. J. V., and Kazazian, H. H., 1982, Non-random association of polymorphic restriction sites in the β-globin gene cluster. *Proc. Natl. Acad. Sci. USA* **79:**137–141.
21. Sokal, D. C., Byrd, J. R., Chen, A. T. L., Goldberg, M. F., and Oakley, G. P., Jr., 1980, Prenatal chromosomal diagnosis: Racial and geographic variation for older women in Georgia, *J. Am. Med. Assoc.* **244:**1355–1357.

# 4

# Economic Impact of Recombinant DNA Technology

J. LESLIE GLICK

## INTRODUCTION

Recombinant DNA technology (RDT) is a generic technology. It can be applied to the development of a wide range of product lines in diverse industrial settings. It enables one to design microbes with genetic capabilities to form products which otherwise could not be synthesized by means of fermentation.

There are two main approaches in applying RDT. The first is to modify genetically those microorganisms which are already capable of forming commercially desirable products. Not only is it possible to introduce specific structural and regulator genes into a bacterium, it is also feasible to permit the desired genes to be amplified by as many as several hundred copies per bacterium. The amount of product coded by the foreign gene is roughly proportional to the number of copies of that gene in the host bacterium. Genetic modification of such microbes results in vastly improved yields of natural products. Greater biosynthetic efficiencies result in significant cost reductions with concomitant increases not only in gross margins but also in revenues for those products whose demand is limited due to excessively high costs of production.

Genetically improved microbial processes currently utilized in the fermentation industry include those involved in the production of amino acids, enzymes, antibiotics, and steroids. Genetic manipula-

J. LESLIE GLICK • Genex Corporation, Rockville, Maryland 20852.

tion can also be applied to microorganisms which are naturally capable of synthesizing commercially desirable products but at economically unattractive yields. Such products include pesticides, fuels, and industrial cyclic and acyclic chemicals. Some of these are complex compounds which are synthesized chemically from high-cost raw materials in multistep reactions but are also produced in microorganisms in trace amounts. Genetic engineering and application of fermentation technology may thus result in economical alternatives to current modes of manufacture.

The second approach in applying RDT is to design microorganisms which normally would not form such products as vaccines and hormones. Microbial strains can be constructed with genes isolated from foreign sources, e.g., hepatitis virus, in order to enable bacteria to synthesize hepatitis viral antigens for use as a vaccine. Strains can also be constructed with genes synthesized chemically, e.g., the gene which codes for the hormone insulin. The purpose of this chapter is to consider some of the far-reaching applications of RDT and the potential economic consequences of applying this technology.

## APPLICATIONS

Tables I to V summarize both present and potential applications of RDT to five major industries: pharmaceuticals, agriculture, food processing, chemicals, and energy. The tables distribute 15 different classes of products and their corresponding major uses throughout the five industrial settings. The product categories range from such well-defined product lines as amino acids to such highly speculative entries as gene preparations.

RDT is currently being applied to the development of microbial strains for production of amino acids in high yield. It has already been utilized to enhance significantly the levels of the enzymes involved in tryptophan[1] and isoleucine[2] syntheses. RDT will have a particularly large impact on those essential amino acids which are expensive to manufacture. For example, tryptophan and threonine, which cost as much as \$40–\$50 per pound, would be used extensively as feed additives to corn if the cost could be reduced to \$2–\$5 per pound. By way

**Table I.** Applications of Recombinant DNA Technology to the Pharmaceutical Industry

| Product category | End use |
|---|---|
| Amino acids | Intravenous solutions and therapeutics |
| Vitamins | Intravenous solutions and prophylactics |
| Petrochemicals | Analgesics, e.g., aspirin and acetaminophen |
| Steroid hormones | Prophylactics and therapeutics, e.g., anti-inflammatory agents |
| Antibiotics | Control of infectious diseases |
| Short peptides | Manufacturing processes, e.g., in tissue culture |
| Peptide hormones | Control of metabolic disorders, e.g., insulin and human growth hormone |
| Enzymes | Diagnostics, e.g., glucose oxidase; and therapeutics, e.g., urokinase |
| Viral antigens | Vaccines |
| Other proteins | Therapeutics, e.g., interferon and human serum albumin |
| Gene preparations | Control of hereditary disorders, e.g., sickle cell anemia |

of comparison one should look at the marketability of methionine and lysine, which are sold in the range of $1–$2 per pound and are widely used as feed additives. Worldwide sales of methionine and lysine together exceed $400 million, in contrast to sales of tryptophan and threonine which amount to around $25 million.

**Table II.** Applications of Recombinant DNA Technology to the Agricultural Industry

| Product category | End use |
|---|---|
| Amino acids | Feed additives |
| Vitamins | Feed additives |
| Petrochemicals | Feed preservatives, e.g., sorbic acid |
| Antibiotics | Feed additives and prophylactics |
| Peptide hormones | Growth promoters, e.g., bovine growth hormone |
| Viral antigens | Vaccines |
| Enzymes | Feed additives, e.g., papain; and feed preservatives, e.g., glucose oxidase |
| Pesticides | Insecticides, e.g., bacilli spores |
| Gases | Fertilizer, e.g., due to ammonia |

**Table III.** Applications of Recombinant DNA Technology to the Food-Processing Industry

| Product category | End use |
|---|---|
| Amino acids | Food enrichment and flavoring agents |
| Vitamins | Food enrichment |
| Petrochemicals | Food preservatives, e.g., benzoic and propionic acids |
| Short peptides | Artificial sweeteners, e.g., aspartame |
| Enzymes | Manufacturing processes, e.g., amylases and glucose isomerase |
| Nucleotides | Flavoring agents |

**Table IV.** Applications of Recombinant DNA Technology to the Chemical Industry

| Product category | End use |
|---|---|
| Petrochemicals | Cyclic and acyclic intermediates and end products |
| Gases | Manufacturing processes, e.g., ammonia and hydrogen |
| Enzymes | Manufacturing processes, e.g., proteases |
| Minerals | Catalysts and inorganic intermediates, e.g., transition metals |

**Table V.** Applications of Recombinant DNA Technology to the Energy Industry

| Product category | End use |
|---|---|
| Enzymes | Manufacturing processes, e.g., ethanol dehydrogenase and hydrogenase |
| Petrochemicals | Fuel, e.g., ethanol |
| Gases | Fuel, e.g., methane and hydrogen |
| Minerals | Fuel, e.g., uranium |

**Table VI.** Effect of Large-Scale Production of Methionine on Its Bulk Selling Price and Total Sales

| Year | World volume (million lb) | Bulk selling price ($/lb)[a] | World market ($ millions)[a] |
|---|---|---|---|
| 1954 | 3 | 8.30 | 25 |
| 1960 | 10 | 3.50 | 35 |
| 1970 | 60 | 2.21 | 133 |
| 1980 | 210 | 1.20 | 252 |

[a]In 1980 dollars.

Tables VI and VII demonstrate the inverse relationship between the bulk selling prices and the amounts of methionine and lysine produced over a period of two decades. Table VII shows the effect of production efficiency on the bulk selling price of lysine. As yields improved, the cost of manufacture dropped. The economic availability of lysine coupled with the growing demand for it stimulated production enormously. This resulted in greatly increased sales revenues despite the marked lowering of the unit selling price. Today lysine is produced in yields as high as 80–100 g/liter, in contrast to tryptophan, which is produced in yields of only 10–15 g/liter. Clearly improvement of tryptophan yields by means of RDT would lower the cost of manufacture and thereby stimulate demand for this important amino acid.

In general RDT has immediate applications to the improvement of yields of microbial enzymes. For example, a 500-fold increase in activity of DNA ligase resulted following manipulation of *E. coli* by

**Table VII.** Impact of Improved Technology on the Lysine Market

| Year | Production efficiency (arbitrary units) | Bulk selling price ($/lb)[a] | World volume (million lb) | World market ($ millions)[a] |
|---|---|---|---|---|
| 1958 | 1 | 23.70 | 0.2 | 5 |
| 1969 | 4 | 4.45 | 7.0 | 31 |
| 1980 | 8 | 2.10 | 90.0 | 189 |

[a]In 1980 dollars.

means of RDT.[3] While DNA ligase is not of much commercial significance (although it is a necessary tool of RDT), a variety of other enzymes are of extreme importance. A number of manufacturing processes in major industries are absolutely dependent on the action of these enzymes, whose availability controls several billion dollars' worth of business.

Glucose isomerase, the enzyme which converts glucose to fructose, not only has worldwide sales of $30–$40 million but is now an essential factor in the multi-billion-dollar sweetener market. Fructose is sweeter than sucrose, which is sweeter than glucose. Nevertheless high-fructose corn syrup (HFCS), which is a mixture of glucose and fructose, is still sweeter than sucrose with fewer calories for the same degree of sweetness. Less expensive than sucrose, HFCS is gaining wide acceptance as an alternative to sucrose.

While over 1000 enzymes have been characterized and purified, only about 50 have any commercial significance. Of these 50, sales of microbial proteases and amylases alone represent over half the enzyme market. However, many other enzymes, including plant and animal enzymes, would be distributed more extensively for commercial application if their costs of production could be sharply reduced. Here RDT offers a unique opportunity. For example, the calf prorennin gene has already been cloned in *E. coli*.[4] Prorennin is the proenzyme of rennin, which is important for cheese production. As another example, Abbott Laboratories recently announced that it had been able to produce human urokinase in *E. coli*.[5] An extremely expensive therapeutic (a couple of thousand dollars per dose), urokinase would be used more frequently to dissolve acute pulmonary embolisms if it were cheaper to isolate. Production of this enzyme by means of fermentation should eventually lower its cost at least 90%–99%.

Clearly one application of RDT which is already well underway toward commercialization is the production of peptide hormones in *E. coli*. Human insulin[6] and growth hormone[7,8] genes were reported to be expressed in *E. coli* in 1979. The following year Eli Lilly & Co. announced that it had purified and crystallized microbially derived insulin with biological activity identical to that of pancreatic insulin.[9] The initial results of clinical studies were sufficiently encouraging that

Eli Lilly committed $40 million toward construction of fermentation facilities for large-scale manufacture of human insulin. With respect to microbially produced growth hormone, there is an especially pressing need for it. Most individuals suffering from hypopituitary dwarfism cannot afford to purchase the extraordinarily expensive, and scarce, naturally occurring hormone. Eventually there should even be a sizable market for microbially produced animal growth hormones, such as bovine growth hormone, with potential applications to the meat and dairy industries.

Another area well under attack by RDT involves the production of vaccines, which either are not readily produced or are relatively ineffective when produced by means of conventional technology. Hepatitis B virus (HBV) antigens have already been expressed in *E. coli.*[10–12] In fact a plasmid containing the HBV genome has even been used to infect cultured mouse fibroblasts so as to enable them to synthesize and excrete HBV surface antigen particles.[13] Influenza antigens have also been expressed in *E. coli.*[14] Genentech and Plum Island Animal Disease Center are currently undertaking a collaborative project to develop a vaccine against foot and mouth disease virus (FMDV) by cloning segments of the FMDV genome.[15]

Recently much attention has focused on the cloning and expression of biologically active leukocyte[16–18] and fibroblast[19] interferons in *E. coli.* Microbially produced human interferon can already be produced economically with efficacy comparable to that of naturally occurring interferon in protecting squirrel monkeys from lethal encephalomyocarditis viral infection.[18] At least four genetic engineering firms—Biogen, Cetus, Genentech, and Genex—are extensively involved in interferon projects, and at least five major corporations—Bristol-Myers, du Pont, Hoffmann-LaRoche, Schering-Plough, and Shell Oil—have entered the race for commercial development of microbially produced interferon.

Many kinds of biologically active peptides will eventually be produced in genetically engineered microorganisms. They include endorphins, clotting factors, tissue growth factors, and perhaps even monoclonal antibodies. A biologically active β-endorphin was recently cloned and expressed in *E. coli.*[20] Genentech has announced that it has produced thymosin from a genetically engineered microorganism.[21] However, the most immediate application of RDT for this group of products may involve the production of human serum albumin. Currently prepared as a crude plasma fraction, human serum albumin is relatively expensive because of its limited supply. Much in demand for treatment of shock, hypoproteinemia, and erythroblastosis fetalis, human serum albumin currently commands a world market of perhaps $250 million. Ovalbumin, a related avian protein, was one of the first eukaryotic proteins to be expressed at high levels in *E. coli*.[22,23]

Other areas in which RDT should play an important role for the pharmaceutical industry include steroids and antibiotics. In the past 15 years production of steroid hormones has increased approximately fourfold. What has been exceedingly important for the rapid growth of the steroid market is the application of microbial enzymes in lowering the costs of manufacture. For example, in 1949 cortisone cost $200 per gram because of its very complicated and inefficient method of synthesis. A few years later the price dropped to $6 per gram because of the discovery of a microbial enzyme which played a key role in a much more economic method of synthesis. Subsequent improvements even permitted the price to drop further, to under 50¢ per gram in 1980.

With respect to antibiotics, we still lack a solid understanding of the genetics which control the various biosynthetic pathways. Nevertheless more efficient yields of antibiotics may result from genetic manipulation of those microbes which produce a particular enzyme that is capable of modifying a known antibiotic or an inactive precursor to a more effective form. Examples of enzymatic processes that result in antibiotic end products or important intermediates include the conversion of benzylpenicillin to 6-aminopenicillanic acid (6-APA), cis-propenylphosphonic acid to fosfomycin, novobiocin to novenamine, 12a-deoxytetracycline to tetracycline, 5a,11a-dehydrochlortetracycline to 5-hydroxy-7-chlortetracycline, and daunomycin to adriamycin. Recently the gene for penicillin G acylase was cloned onto multicopy plasmids in *E. coli*, thereby causing a 50-fold increase in the level of acylase activity and permitting production of 6-APA in an immobilized cell reactor.[24] However, RDT is only just beginning to be applied to enhance yields of antibiotics. It has only been a couple of years since a practical means was devised to permit transformation of *Streptomyces* by plasmid DNA.[25] Over 60% of known antibiotics are produced by *Streptomyces*. More recently a DNA cloning system was devised for transferring genes between species into antibiotic-producing *Steptomyces*.[26]

The farthest-reaching applications of RDT to the pharmaceutical industry undoubtedly will extend to gene therapy. Sickle cell anemia and Cooley's anemia represent those hereditary disorders most likely to be treatable by purified gene preparations. Probably over two million persons worldwide suffer from these two disorders alone. In 1979 Mulligan *et al.*[27] first demonstrated that the rabbit β-globin gene could be inserted into monkey kidney cells and direct synthesis of rabbit β-globin. Other laboratories subsequently demonstrated that the rabbit β-globin gene could be transferred to and expressed in mouse fibroblasts.[28,29] The human β-globin gene has also been transferred to mouse cells.[30] More recently gene transfer into murine bone marrow cells was conclusively demonstrated *in vivo*.[31,32] The bone marrow of mice injected with the DNA-treated cells was populated primarily with cells transformed by specific genetic markers.

Proceeding from the edge of the technology back to some potentially large-scale applications of RDT, let us return to the food-processing industry. This industry has long practiced the art of fermentation. Wine, beer, cheese, pickles, vinegar, sauerkraut, and mushrooms are all produced due to the action of microorganisms. Advances in RDT have an obvious role to play in the improvement of those microbial processes currently involved.[33] The importance of genetic engineering was stressed in the previous discussion on the production of amino acids for use as feed additives. However, about half of the world market for amino acids is attributed to glutamic acid, due to its use as the flavoring agent monosodium glutamate (MSG). The 5′-nucleotides, inosinic acid and guanylic acid, are also used as flavoring agents. By means of RDT the dosage of genes coding for the enzymes of the nucleotide pathways could be increased by placing the pertinent genes on a multicopy plasmid. Insertion of such plasmids into a microbial strain containing appropriate regulatory mutations should enable the organism to enhance its yield of nucleotides.

Earlier we considered the importance of glucose isomerase in providing the less caloric sucrose alternative HFCS. One application of RDT, which may be especially important to the artificial sweetener market, pertains to the production of the practically noncaloric aspartame. The C-terminal methyl ester of the aspartylphenylalanine dipeptide, aspartame is 100–200 times as sweet as sucrose. The world market for aspartame is currently only a few million dollars. However, once its use as a sweetener is permitted by appropriate government regulatory agencies, its market may expand to over $2 billion. While aspartame is now manufactured by chemical means, it could be produced by fermentation at a cost competitive to its cost of chemical synthesis. By means of RDT one can construct a microorganism with a polyaspartylphenylalanine gene that will enable the organism to produce large quantities of a polypeptide consisting entirely of the repeating dipeptide aspartylphenylalanine.[34] The dipeptides may be enzymatically cleaved and methylated to aspartame by chemical or enzymatic means. Alternatively, microorganisms could be genetically engineered to produce aspartic acid and phenylalanine, the precursors of aspartame. The market for both these amino acids should increase enormously once aspartame is widely available.

Another product which should be noted is single-cell protein (SCP). A potential commodity to the feed and food markets, SCP will not be widely utilized until its cost becomes more competitive with that of traditional sources of protein. RDT has already been applied to improving the conversion of methanol to SCP in bacteria.[35,36]

Switching areas again, let us briefly mention the pesticide industry as a target of RDT. Considering just the control of insects, only 5% of this market is served by microbial insecticides while the bulk of it is met by chemicals. It should be noted that bioconversions resulting in chemical insecticides may be achieved by means of microbial enzymes. While RDT has a role to play here, it undoubtedly will focus on the construction of more potent microbial insecticides. Around 100 species of bacteria and several types of viruses are known as pathogens to insects. At least three species of *Bacillus* and two different viruses have been produced commercially.

The potentially largest applications of RDT will likely impact on the production of gases, minerals, and petrochemicals, and on the control of air and water pollution. For example, the Environmental

Protection Agency has estimated that in the U.S. alone between 1977 and 1986 government and industry will have spent $360 billion on pollution control.[37] Of that amount, between $20 and $25 billion will have been spent by chemical-related industries. Microbial processes are currently being developed to treat a variety of wastes, including phenols, arylsulfonates, organic phosphates, and heavy metals.[38] The bioaccumulation of inorganic contaminants may result in the recovery of metals such as uranium, copper, and zinc, whereas the biotransformation of organic wastes can lead to the formation of fuels such as methane, hydrogen, and ethanol. Novel uses of genetically engineered microorganisms should eventually lead to the photosynthetic production of hydrogen from water,[39] leaching of minerals from low-grade ores,[40] desulfurization of petroleum and coal,[41] and enhanced recovery of oil from marginally producing wells.[42]

Many chemical intermediates are increasingly rising in price because of the soaring costs of petroleum, both as a fuel and as a raw material. Fortunately, commodity petrochemicals as well as specialty compounds are produced naturally by microorganisms, although in minuscule amounts.[43] Genetic engineering of such strains, coupled with novel approaches to fermentation technology, should lead to large-scale manufacture of chemicals by means of microbial processes. Cyclic compounds which may eventually be produced biologically include phenol, aniline, benzoic acid, salicyclic acid, and acetaminophen. Acyclics which may be produced similarly include ethylene and propylene glycols; acetic, acrylic, and adipic acids; and sorbitol and sorbic acid.

For example, Cetus has designed a multi-enzyme process which can convert ethylene and propylene to their corresponding oxides.[44] The process requires hydrogen peroxide, which can be generated enzymatically during the oxidation of glucose. If the glucose is oxidized to glucosone, the latter could be hydrogenated chemically to form fructose. Virtually all of the glucose would be converted to fructose, unlike the conventional glucose isomerase process where only half the glucose is converted to form HFCS.

Another example of a cost-effective biological production of a commodity chemical would be the manufacture of ethanol by immobilized, genetically engineered thermophilic organisms. Genex has an interest in the application of RDT to construct strains capable of fermentation at temperatures in excess of 80°C. The strains would be utilized in a process that combines fermentation and distillation in a single reactor. The advantages include unusually high fermentation rates, direct formation of ethanol vapor, virtual absence of inhibition of fermentation by ethanol, reduced energy costs from the minimal cooling requirements, and almost no risk of contamination. We estimate that the cost of manufacturing ethanol by means of this biological process would truly be competitive with the cost of manufacturing ethanol chemically from ethylene.

The impact of genetic engineering on the chemical industry will be overwhelming only when microorganisms are designed to recycle waste streams effectively and to utilize efficiently such plant-derived polymers as cellulose and lignin. Eventually new substances will be produced biologically to substitute for existing products that are synthesized chemically from ethylene monomers. The plastics industry may be profoundly affected. Naturally occurring, microbially produced polymers such as pullulan (a product of maltotriose subunits) and poly-β-hydroxybutyrate represent nature's answer to plastics. Pullulan, for example, can be thermally processed without the need of a plasticizer to resemble a transparent, commercial food wrap. It may be used in a variety of applications, such as a packaging film, textile yarn, gauze, adhesive, or even a molded product. A major advantage of pullulan is that it is not a polluting material. It is nontoxic and biodegradable. The commercial production of pullulan represents a major challenge and opportunity for an interdisciplinary team of genetic and biochemical engineers.

## ECONOMICS

Table VIII projects the impact of RDT on world markets in the year 2000. The 15 product categories listed here correspond to the product categories that are distributed throughout Tables I to V. The reader should keep in mind the following six caveats before considering the numbers in this and subsequent tables.

First, one must assume 20-year projections to be inaccurate at

**Table VIII.** Impact of Recombinant DNA Technology on Bulk Sales in the Year 2000

| | | Market values ($ millions)[a] | | | |
|---|---|---|---|---|---|
| Product category | Number of compounds | Current | Portion of current[b] | Growth[c] | Total[d] |
| Amino acids | 20 | 1,700 | 1,700 | 700 | 2,400 |
| Vitamins | 10 | 680 | 430 | 0 | 430 |
| Enzymes | 50 | 300 | 200 | 300 | 500 |
| Steroid hormones | 20 | 380 | 130 | 0 | 130 |
| Peptide hormones | 10 | 260 | 260 | 1,000 | 1,260 |
| Viral antigens | 15 | 0 | 0 | 200 | 200 |
| Short peptides | 2 | 10 | 10 | 2,000 | 2,010 |
| Nucleotides | 2 | 70 | 70 | 0 | 70 |
| Misc. proteins | 5 | 300 | 300 | 1,200 | 1,500 |
| Antibiotics | 80 | 4,200 | 2,000 | 0 | 2,000 |
| Gene preparations | 3 | 0 | 0 | 100 | 100 |
| Pesticides | 5 | 100 | 100 | 100 | 200 |
| Petrochemicals | 270 | 36,000 | 7,000 | 18,000 | 25,000 |
| Gases | 3 | 44,000 | 0 | 2,400 | 2,400 |
| Minerals | 5 | 37,000 | 1,800 | 0 | 1,800 |
| Total | 500 | 125,000 | 14,000 | 26,000 | 40,000 |

[a] World market values in 1980 dollars of products susceptible to applications of recombinant DNA technology.
[b] Estimated portion of current (1980) market on which recombinant DNA technology is projected to impact in 20 years.
[c] Incremental growth of market in 20 years due to impact of recombinant DNA technology.
[d] Total = Portion of current + Growth = Projected market in 20 years for products to which recombinant DNA technology will be applied.

best and outrageous at worst. In this case we are looking at the impact of a brand-new technology. For practical purposes, it has contributed to zero sales worldwide in 1980.

Second, advances in RDT sometimes seem to proceed with the speed of light. The problem here is that normally one should prudently temper the highly speculative nature of 20-year projections. Yet if RDT were to grow as rapidly as semiconductor technology, a conservative approach to 20-year projections could be totally meaningless. Sales of one semiconductor company, for instance, jumped from $8 million in 1971 to $660 million only eight years later.[45] In 1977 I polled a number of molecular biologists, most of whom predicted that the first molecules of insulin and interferon would not be micro-bally produced in the research laboratory before 1982 and 1987, respectively. These two events occured in 1978 and 1979, respectively.

Third, Table VIII does not attempt to project the impact of RDT on every conceivable product. In fact these projections exclude what could be gigantic opportunities. Single-cell protein is not considered here, nor is the whole area of biological control of air and water pollution. Further, many more enzymes, hormones, and antibiotics exist than are enumerated in the table. Undoubtedly new products and applications will occur in the next 20 years which play no role in these projections. Table VIII also excludes consideration of foods such as wine, beer, cheese, vinegar, or bakers' yeast. Even HFCS is excluded, because glucose isomerase, the enzyme responsible for its production, is included in the enzyme product category. Thus what may be overoptimism in some areas is offset with excessive discounting in others.

Fourth, all projections are based on zero inflation. Therefore all dollar values are based on 1980 dollars.

Fifth, all projections are based on zero population growth.

Sixth, current market values of the product lines listed in Table VIII should not be treated as hard data. There is no single reliable source of market data. Individual companies tend to be secretive about sales of individual products. Nevertheless a reasonable financial picture can be pieced together by incorporating a fair amount of census data with information available in corporate annual reports and technology review articles. This data base is supplemented by various disclosures made by industrial sources, trade organizations, investment bankers, and independent consultants.

The current market values of 500 compounds considered as potential targets of RDT total $125 billion (Table VIII). This sum represents bulk sales or items sold wholesale. It is probably accurate to within 20%. Its largest components—the market values of selected gases, minerals, and petrochemicals—represent the hardest data. These three product categories alone account for $117 billion, or almost 94% of the total. If anything, the values used here are too low because they reflect figures as old as four years. Market values of the three gases considered—methane, hydrogen, and ammonia—were the easiest to obtain.[46,47] However, the value of methane (90% of the value of all three gases) was based on data available for 1976. The price of methane has risen considerably since then. Market values of the five minerals—cobalt, nickel, manganese, iron, and uranium—were also readily calculated,[46,48] as were the market values of those petrochemicals selected as targets of biotechnology.[46,47,49] Publicly available information was particularly useful in estimating market values of amino acids[47,50,51] and enzymes[52–54] but was of lesser importance in arriving at market values of vitamins,[47,49] steroid hormones,[47,55] peptide hormones,[9,54] nucleotides,[56] antibiotics,[55,57] and pesticides.[58] For these latter categories input from confidential sources was especially helpful. Market values of the remaining product lines listed in Table VIII were estimated only from proprietary information.

If determining the size of a market is like seeing at night, predicting its future is like seeing at night blindfolded. The projections in Table VIII are based on an ongoing technology assessment and market research program initiated at Genex in 1977. Table VIII attempts to split the forecast into two groups: (1) products with relatively mature markets, which through application of RDT might be manufactured at lowered costs; and (2) products with limited or virtually no market because demand is restricted by excessive manufacturing costs.

It is assumed that there is a small degree of elasticity to the markets corresponding to the first group of products. That is, decreases in selling price would tend to stimulate demand for the products. However, revenues would remain relatively constant. Market size would appreciate somewhat, but the resulting increase in sales would be offset by lower selling prices. The drop in selling prices would reflect a decrease in manufacturing costs because of the competition between new and old technology. Such competition might lead to redistribution of market share among the various manufacturers.

The second group of products, while including such speculative possibilities as interferon, pullulan, and gene preparations, also contains some commodity items. For example, an efficient means of producing ethanol biologically would partially satisfy a growing need for an alternative automative fuel. Yet such production would not necessarily compete with conventional technology regarding the manufacture of ethanol for industrial use. As another example, microbial production of gases would probably not be competitive enough in 20 years to even begin to displace conventional means of production. However, the new technology might be sufficiently cost-effective to supplement conventional technology under certain circumstances, such as in the formation of methane from organic wastes.

As projected in Table VIII, RDT may impact on $40 billion worth of products in the year 2000. Of this total, $14 billion applies to old markets and $26 billion applies to new markets. Clearly the largest portion of the $40 billion, 62.5% or $25 billion, affects the petrochemical industry.

Table IX summarizes the projected impact of RDT on corporate earnings worldwide in the year 2000. (A simplifying but inaccurate assumption is made here, namely, that net income after taxes is always one-half of pretax income.) In only 20 years annual earnings from products produced as a result of RDT may reach $3.5 billion, of which $1.8 billion would flow to the petrochemical industry.

Table X estimates the size of corporate investment required to realize the above scenario for commercial application of RDT. Over a 20-year period this investment totals $24 billion, of which $20 billion pertains to plant capital expenditures and $4 billion to research and development. However, the bulk of this investment, 79% or $19 billion, is required to apply biotechnology to the petrochemical indus-

**Table IX.** Impact of Recombinant DNA Technology on Corporate Earnings in the Year 2000

| | Net income ($ millions)[a] | | |
|---|---|---|---|
| Product category | Redistribution of current[b] | Growth[c] | Total |
| Petrochemicals | 500 | 1300 | 1800 |
| All other products | 700 | 1000 | 1700 |
| Total | 1200 | 2300 | 3500 |

[a]Worldwide net income after taxes in 1980 dollars.
[b]Estimated redistribution of net income on current sales of products due to impact of recombinant DNA technology.
[c]Incremental growth of income due to growth of market caused by impact of recombinant DNA technology.

**Table X.** Commercial Investment Required for Optimal Impact of Recombinant DNA Technology in the Year 2000

| | Investment ($ millions)[a] | | | | |
|---|---|---|---|---|---|
| Product category | Lab R&D[b] | Pilot plant R&D[b] | Licensing[c] | Plant capital expenditures[d] | Total |
| Petrochemicals | 500 | 1,600 | 300 | 16,600 | 19,000 |
| All other products | 300 | 300 | 1,000 | 3,400 | 5,000 |
| Total | 800 | 1,900 | 1,300 | 20,000 | 24,000 |

[a]Cumulative 20-year investment in 1980 dollars which is required to apply recombinant DNA technology in order to achieve sales projections for the year 2000.
[b]Capital expenditures for lab and pilot plant are excluded, except for treatment of depreciation on individual projects.
[c]Licensing pertains to compliance with government regulations and includes animal and clinical studies, where appropriate.
[d]Plant capital expenditures are minimized by utilization of existing facilities wherever possible.

try. (Here a simplifying assumption is made that understates the degree of investment necessary for achieving the revenues projected in Table VIII. There is little attempt to determine the extent of competition in research and development projects. Competitive efforts will lead to duplication of investment, particularly in the research and development stages.)

Table XI analyzes the projected return on investment to the private sector. The percent return is presented two ways. The first de-

**Table XI.** Analysis of Corporate Return on Investment Pertaining to Recombinant DNA Technology in the Year 2000

| | Percent return on investment | |
|---|---|---|
| Product category | Plant capital expenditures[a] | Plant capital expenditures and R&D[a] |
| Petrochemicals | 10.8 | 9.5 |
| All other products | 50.0 | 34.0 |
| Total | 17.5 | 14.6 |

[a]These percentages were obtained by dividing the total net income projections in Table IX by the appropriate investment projections in Table X.

rives percent return without considering research and development investment. This approach, which yields a return of 17.5%, only treats plant capital expenditures as investment. The second approach includes research and development expenditures as part of the overall investment and lowers the return to 14.6%.

The cost of research and development has relatively little effect on return on investment with respect to petrochemicals, reducing the projected return from 10.8% to 9.5%. In contrast, the cost of research and development, when applied to other product categories, reduces the projected return on investment significantly, from 50% to 34%. Yet the return on investment for these other products is still projected almost four times higher than for the petrochemicals. As can be gleaned from Table X, over 87% of the investment required to apply RDT to the petrochemical industry pertains to capital expenditures for plant equipment and facilities. This is not the case for other applications of RDT, which are much less capital intensive. In those situations only 68% of the total investment pertains to capital expenditures for plant equipment and facilities.

The difference between the petrochemical industry and other industries with respect to return on investment should diminish over time. In 20 years many nonpetrochemical product lines resulting from applications of RDT will have been manufactured for over a decade; in contrast, the petrochemical industry will only begin to be affected by RDT. The next 20 years represents too short a period to expect an exciting return on investment for application of RDT to the petrochemical industry. Yet this investment must be made. Feedstocks derived from nonrenewable resources, which are currently vital to the chemical industry, will become increasingly scarce. Moreover, organic chemical reactions, unlike biochemical reactions, typically require conditions of high temperature and high pressure in order to proceed to completion. Low-energy-requiring microbial processes will thus have to be devised to utilize feedstocks from renewable resources. The investment will be huge but so will the return, given sufficient time for the technology to develop and production to scale up.

Table XII attempts to project worldwide the financial return to the public in the year 2000 on bulk sales of products whose manufac-

**Table XII.** Financial Return to the Public in the Year 2000 on Bulk Sales of Products Whose Manufacture Would Be Affected by Recombinant DNA Technology

| Product category | Financial return ($ millions)[a] Cost savings on current products | Taxes due to market growth | Total |
|---|---|---|---|
| Petrochemicals | 1000 | 1300 | 2300 |
| All other products | 1400 | 1000 | 2400 |
| Total | 2400 | 2300 | 4700 |

[a]Based on 1980 dollars.

ture will have been affected by RDT. The financial return of $4.7 billion is achieved two ways. First, the combination of new technology and corporate competition should lead to considerable cost savings on existing product lines ($2.4 billion). Second, manufacture of new products and development of mature markets should provide a source of sizable tax revenues ($2.3 billion). (It is assumed that tax revenues on existing product lines will not fall off as selling prices drop, since expansion of market size would compensate because of increased demand for cheaper products.)

Table XIII estimates the amount of government funds previously committed for recombinant DNA research and projects the amount to be spent in the next decade. While figures are readily available for research funds spent on projects involving recombinant DNA,[48,59–61] these figures must be discounted significantly if one wishes to estimate how much was spent specifically on recombinant DNA research. U.S. government figures do not distinguish projects in which RDT was used as a peripheral methodology from those which focused exclusively on the development of RDT. A rule of thumb suggested by the government agencies is to apply a discount factor of 50%.

In the past decade relatively little was spent on recombinant DNA research in countries other than the U.S. This is rapidly changing. However, trying to predict the level of government spending in the next 10 years, both in the U.S. and abroad, is probably a futile task. Nevertheless Table XIII projects an optimistic level of govern-

**Table XIII.** Public Investment Required for Optimal Impact of Recombinant DNA Technology in the Year 2000

| | ($ millions) | | |
|---|---|---|---|
| Source of research funds | Funds committed in the 1970s for research containing recombinant DNA methodologies[a,b] | Estimated funds committed in the 1970s solely for recombinant DNA research[a] | Cumulative funds spent in the 1970s and projected for the 1980s solely on recombinant DNA research[c] |
| United States | | | |
| NIH | 269 | 135 | |
| NSF | 33 | 16 | |
| Other | 4 | 2 | |
| Subtotal | 306 | 153 | 1500 |
| Other nations (estimated) | 60 | 30 | 1500 |
| Total | 366 | 183 | 3000 |

[a]Includes funds spent through mid-1980.
[b]Includes funds for research projects that involve research besides recombinant DNA.
[c]All future funds are based on 1980 dollars.

ment spending on recombinant DNA research, totaling $3 billion for the two decades ending 1990. Half of this amount would be spent by the U.S. government. Projections are not attempted for the 1990s since it is assumed that the lead time for application of basic research findings and subsequent commercialization is of the order of 10 years. All of the other projections are based on the year 2000.

Table XIV compares the projected return on investment pertaining to RDT to the private and public sectors in the year 2000. The 14.6% return on investment to the private sector represents projected corporate earnings of $3.5 billion (Table IX) divided by projected corporate investment of $24 billion (Table X). The 156.7% return on investment to the public represents the projected financial return of $4.7 billion (Table XII) divided by projected cumulative government spending of $3 billion (Table XIII). What is most striking is that the public may benefit extraordinarily well compared to the private sector, realizing a tenfold higher percentage return on an investment one-eighth that of the private sector. Government spending on recom-

**Table XIV.** Comparison of Private and Public Return on Investment Pertaining to Recombinant DNA Technology in the Year 2000

| Sector | Percent return on investment |
|---|---|
| Private | 14.6 |
| Public | 156.7 |

binant DNA research may well represent one of the best investments the public could make. Clearly this is one example where both the public and the private sector will benefit from a strong, government-funded basic research program.

## CONCLUSIONS

I have painted a very bright picture for RDT and have tried to point out the underlying assumptions for this scenario. Factors were discussed which could impact on the above projections, both favorably and adversely. Nevertheless with all of the documentation presented one should still take these projections with a grain of salt.

It appears quite evident that this exciting technology will be brought to fruition because of the combined efforts of government, the universities, and industry. For such collaboration to work each party should do what each does best. Government-sponsored basic research must flourish at the universities and other research centers. Basic research findings represent the underpinnings of technology. Industry is quickly applying these findings to an enormous range of problems in the areas of health, agriculture, energy, and the environment.

One cautionary note that might be mentioned here pertains to the alarming trend for universities to become perhaps overzealous in attempting to subject basic research findings to patent protection. There are even instances where universities are considering applying their technological resources in business ventures. Excessive secrecy will wither the scientific grapevine, for only after patents are awarded will researchers feel free to communicate with one another. Worse, academic scientists may find their research findings locked up as trade secrets. Contrary to what university administrators believe, such aggressive patent and investment policy may kill the Golden Goose. Basic research will suffer and therefore so will applied research. Universities are hard pressed for funds, but their latest schemes to get rich may ultimately lead the government to turn off the money spout. It is in the public's best interest that universities receive ample support for basic research from both government and industry, and leave the commercial side to industry.

## REFERENCES

1. Hershfield, V., Boyer, H. W., Yanofsky, C., Lovett, M. A., and Helinski, D. R., 1974, Plasmid ColEl as a molecular vehicle for cloning and amplification of DNA, *Proc. Natl. Acad. Sci. USA* **71:**3455–3459.
2. Davis, M. G., and Calvo, J. M., 1977, Relationship between messenger ribonucleic acid and enzyme levels specified by the leucine operon of *Escherichia coli* K-12, *J. Bacteriol.* **131:**997–1007.
3. Panasenko, S. M., Cameron, J. R., Davis, R. W., and Lehman, I. R., 1977, Five hundredfold overproduction of DNA ligase after induction of a hybrid lambda lysogen constructed *in vivo, Science* **196:**188–189.
4. Beppu, T., Nishimori, K., Kawaguchi, Y., Uchiyama, H., and Uozumi, T., 1980, Cloning of cDNA of calf prorennin in *Escherichia coli, VI International Fermentation Symposium,* London, Ontario, Canada, Abstract, p. 179.
5. Clot dissolving enzyme made by gene splicing, 1980, *Chem. Eng. News,* April 14, p. 6.
6. Goeddel, D. V., Kleid, D. G., Bolivar, F., Heyneker, H. L., Yansura, D. G., Crea, R., Hirose, T., Kraszewski, A., Itakura, K., and Riggs, A. D., 1979, Expression in *Escherichia coli* of chemically synthesized genes for human insulin, *Proc. Natl. Acad. Sci. USA* **76:**106–110.
7. Goeddel, D. V., Heyneker, H. L., Hozumi, T., Arentzen, R., Itakura, K., Yansura, D. G., Ross, M. J., Miozzari, G., Crea, R., and Seeburg, P. H., 1979, Direct expression in *Escherichia coli* of a DNA sequence coding for human growth hormone, *Nature* **281:**544–548.
8. Martial, J. A., Hallewell, R. A., Baxter, J. D., and Goodman, H. M., 1979, Human growth hormones: Complementary DNA cloning and expression in bacteria, *Science* **205:**602–607.
9. Burnett, J. P., 1980, Biosynthetic human insulin, Presentation at the Second Annual Biotechnology Investment Seminar, sponsored by E. F. Hutton, Inc., New York (September 17, 1980).

10. Burrell, C. J., MacKay, P., Greenaway, P. J., Hofschneider, P. H., and Murray, K., 1979, Expression in *Escherichia coli* of hepatitis B virus DNA sequences cloned in plasmid pBR322, *Nature* **279:**43–47.
11. Pasek, M., Goto, T., Gilbert, W., Zink, B., Schaller, H., MacKay, P., Leadbetter, G., and Murray, K., 1979, Hepatitis B virus genes and their expression in *E. coli, Nature* **282:**575–579.
12. Charnay, P., Gervais, M., Louise, A., Galibert, F., and Tiollais, P., 1980, Biosynthesis of hepatitis B virus surface antigen in *Escherichia coli, Nature* **286:**893–895.
13. Dubois, M–F., Pourcel, C., Rousset, S., Chany, C., and Tiollais, P., 1980, Excretion of hepatitis B surface antigen particles from mouse cells transformed with cloned viral DNA, *Proc. Natl. Acad. Sci. USA* **77:**4549–4553.
14. Emtage, J. S., Tacon, W. C. A., Catlin, G. H., Jenkins, B., Porter, A. G., and Carey, N. H., 1980, Influenza antigenic determinants are expressed from haemagglutinin genes cloned in *Escherichia coli, Nature* **283:**171–174.
15. *Federal Register* **45** (147):50528 (July 29, 1980).
16. Nagata, S., Taira, H., Hall, A., Johnsrud, L., Streuli, M., Ecsodi, J., Boll, W., Cantell, K., and Weissmann, C., 1980, Synthesis in *E. coli* of a polypeptide with human leukocyte interferon activity, *Nature* **284:**316–320.
17. Masucci, M. G., Szigeti, R., Klein, E., Klein, G., Gruest, J., Montagnier, L., Taira, H., Hall, A., Nagata, S., and Weissmann, C., 1980, Effect of interferon-αl from *E. coli* on some cell functions, *Science* **209:**1431–1435.
18. Goeddel, D. V., Yelverton, E., Ullrich, A., Heyneker, H. L., Miozzari, G., Holmes, W., Seeburg, P. H., Dull, T., May, L., Stebbing, N., Crea, R., Maeda, S., McCandliss, R., Sloma, A., Tabor, J. M., Gross, M., Familletti, P. C., and Pestka, S., 1980, Human leukocyte interferon produced by *E. coli* is biologically active, *Nature* **287:**411–416.
19. Derynck, R., Remaut, E., Saman, E., Stanssens, P., DeClercq, E., Content, J., and Fiers, W., 1980, Expression of human fibroblast interferon gene in *Escherichia coli, Nature* **287:**193–197.
20. Shine, J., Fettes, I., Lan, N. C. Y., Roberts, J. L., and Baxter, J. D., 1980, Expression of cloned β-endorphin gene sequences by *Escherichia coli, Nature* **285:**456–461.
21. Genentech, Inc., Form S-1, Registration Statement under the Securities Act of 1933, Securities and Exchange Commission, Washington, D.C., August 19, 1980.
22. Fraser, T. H., and Bruce, B. J., 1978, Chicken ovalbumin is synthesized and secreted by *Escherichia coli, Proc. Natl. Acad. Sci. USA* **75:**5936–5940.
23. Mercereau-Puijalon, O., Royal, A., Cami, B., Garapin, A., Krust, A., Gannon, F., and Kourilsky, P., 1978, Synthesis of an ovalbumin-like protein by *Escherichia coli* K12 harbouring a recombinant plasmid, *Nature* **275:**505–510.
24. Mayer, H., Jaakkola, P., Schomer, U., Segner, A., Kemmer, T., and Wagner, F., 1980, Construction of an *Escherichia coli* strain with high penicillin G acylase-activity by gene cloning, *VI International Fermentation Symposium*, London, Ontario, Canada, Abstract, p. 179.
25. Bibb, M. J., Ward, J. M., and Hopwood, D. A., 1978, Transformation of plasmid DNA into *Streptomyces* at high frequency, *Nature* **274:**398–400.
26. Bibb, M., Schottel, J. M., and Cohen, S. N., 1980, A DNA cloning system for interspecies gene transfer in antibiotic-producing *Streptomyces, Nature* **284:**526–531.
27. Mulligan, R. C., Howard, B. H., and Berg, P., 1979, Synthesis of rabbit β-globin in

cultured monkey kidney cells following infection with a SV40 β-globin recombinant genome, *Nature* **277:**108–114.
28. Mantei, N., Boll, W., and Weissmann, C., 1979, Rabbit β-globin mRNA production in mouse L cells transformed with cloned rabbit β-globin chromosomal DNA, *Nature* **281:**40–46.
29. Wold, B., Wigler, M., Lacy, E., Maniatis, T., Silverstein, S., and Axel, R., 1979, Introduction and expression of a rabbit β-globin gene in mouse fibroblasts, *Proc. Natl. Acad. Sci. USA* **76:**5684–5688.
30. Huttner, K. M., Scangos, G. A., and Ruddle, F. H., 1979, DNA-mediated gene transfer of a circular plasmid into murine cells, *Proc. Natl. Acad. Sci. USA* **76:**5820–5824.
31. Cline, M. J., Stang, H., Mercola, K., Morse, L., Ruprecht, R., Browne, J., and Salser, W., 1980, Gene transfer in intact animals, *Nature* **284:**422–425.
32. Mercola, K. E., Stang, H. D., Browne, J., Salser, W., and Cline, M. J., 1980, Insertion of a new gene of viral origin into bone marrow cells of mice, *Science* **208:**1033–1035.
33. Sessions on fermented foods, 1980, *VI International Fermentation Symposium,* London, Ontario, Canada, Abstract, pp. 140–149.
34. Doel, M. T., Eaton, M., Cook, E. A., Lewis, H., Patel, T., and Carey, N. H., 1980, The expression in *E. coli* of synthetic repeating polymeric genes coding for Poly[L-aspartyl-L-phenylalanine], *Nucl. Acids Res.* **8:**4575–4592.
35. Gautier, F., 1980, Gene cloning for improvement of single cell production by methanol-utilizing bacteria, *VI International Fermentation Symposium,* London, Ontario, Canada, Abstract, p. 102.
36. Windass, J. D., Worsey, M. J., Pioli, E. M., Pioli, D., Barth, P. T., Atherton, K. T., Dart, E. C., Byrom, D., Powell, K., and Senior, P. J., 1980, Improved conversion of methanol to single-cell protein by *Methylophilus methylotrophus, Nature* **287:**396–401.
37. Chementator, 1979, *Chem. Eng. (NY),* October 22, p. 82.
38. Sessions on biological waste treatment and pollution control, 1980, *VI International Fermentation Symposium,* London, Ontario, Canada, pp. 108–118.
39. Session on hydrogen and methane production, 1980, *VI International Fermentation Symposium,* London, Ontario, Canada, pp. 72–76.
40. Murr, L. E., Torma, A. E., and Brierley, J. A. (eds.), 1978, *Metallurgical Applications of Bacterial Leaching and Related Microbiological Phenomena,* Academic Press, New York.
41. Chementator, 1980, *Chem. Eng. (NY),* September 8, p. 19.
42. Calvert, P., 1980, Industrial biopolymers, *Nature* **287:**388.
43. Laskin, A. I., and Lechevalier, H. A. (eds.), 1974, *Handbook of Microbiology,* Volume 4, CRC Press, Cleveland.
44. Neidleman, S. L., Geigert, J., Amon, W. F., Liu, E., and Wolf, B., 1980, A unique multi-enzyme process for the production of alkene oxides, *VI International Fermentation Symposium,* London, Ontario, Canada, Abstract, p. 178.
45. High technology wave of the future or a market flash in the pan?, 1980, *Business Week,* November 10, pp. 86–98.
46. U.S. Bureau of the Census, *Statistical Abstract of the United States: 1978,* 99th edition, Washington, D.C.
47. *Chem. Mark. Rep.,* March and April issues, 1980.
48. U.S. Department of Energy, personal communication, 1980.

49. United States International Trade Commission, 1979, *Synthetic Organic Chemicals: United States Production and Sales, 1978* (USITC Publication 1001), U.S. Government Printing Office, Washington, D.C.
50. Hirose, Y., Sano, K., and Shibai, H., 1978, Amino acids, *in: Annual Reports on Fermentation Processes,* Volume 2 (D. Perlman, ed.), Academic Press, New York, pp. 155–189.
51. Hirose, Y., and Okada, H., 1979, Microbial production of amino acids, *in: Microbial Technology,* Volume 1, 2nd edition (H. J. Peppler and D. Perlman, eds.), Academic Press, New York, pp. 211–237.
52. Aunstrup, K., Andresen, O., Falch, E. A., and Nielsen, T. K., 1979, Production of microbial enzymes, *in: Microbial Technology,* Volume 1, 2nd edition (H. J. Peppler and D. Perlman, eds.), Academic Press, New York, pp. 281–309.
53. Danehy, J. P., and Wolnak, B. (eds.), 1980, *Enzymes: The Interface Between Technology and Economics,* Marcel Dekker, New York.
54. Smith, H., 1980, Presentation at the Second Annual Biotechnology Investment Seminar, sponsored by E. F. Hutton, Inc., New York (September 17, 1980).
55. Sebek, O. K., 1977, Biotransformation of steroid hormones and antibiotics, *in: Biotechnological Applications of Proteins and Enzymes* (Z. Bohak and N. Sharon, eds.), Academic Press, New York, p. 203.
56. Hirose, Y., Enei, H., and Shibai, H., 1979, Nucleosides and nucleotides, *in: Annual Reports on Fermentation Processes,* Volume 3 (D. Perlman, ed.), Academic Press, New York, pp. 253–274.
57. D. Perlman, 1979, Microbial production of antibiotics, *in: Microbial Technology,* Volume 1, 2nd edition (H. J. Peppler and D. Perlman, eds.), Academic Press, New York, pp. 241–276.
58. Ignoffo, C. M., and Anderson, R. F., 1979, Bioinsecticides, *in: Microbial Technology,* Volume 1, 2nd edition (H. J. Peppler and D. Perlman, eds.), Academic Press, New York, pp. 1–26.
59. National Institutes of Health, personal communication (1980).
60. National Science Foundation, personal communication (1980).
61. U.S. Department of Agriculture, personal communication (1980).

# 5

# Foundations and Their Contributions to the Shaping of Health Policy

LEIGHTON E. CLUFF and DAVID E. ROGERS

For many if not most Americans private foundations are poorly understood entities. It is generally recognized that they often award funds for work in areas of social need or cultural affairs, but how they function, or the role they play in our society, remains dim in the minds of most people. Further, as government has increasingly met social needs, some have questioned whether or not the role of private philanthropy has become insignificant, particularly in regard to health and medical care.

Despite the fact that private foundations now make only a relatively small economic contribution to health affairs, we are convinced that they have a significant role to play in helping to improve people's health and medical care in this country. Although there have been rapidly increasing national expenditures in the field, there is general agreement that the health-care industry is simply not keeping up with the demands and needs for better medical care. It is our view that carefully and imaginatively allocated foundation dollars can meet important national purposes. In contrast to public agencies, health-oriented foundations can put their resources into programs which have the long-term objective of improving health care of the future. Programs which may supply better answers to the unresolved problems in medicine and health have been the principal beneficiaries of foundation awards.

LEIGHTON E. CLUFF and DAVID E. ROGERS • The Robert Wood Johnson Foundation, Princeton, New Jersey 08540.

Clearly a major strength of our nation has been the participation of multiple groups in our public affairs. Without philanthropy as a participating partner in the health field, most future innovations, new ideas, and scientific research would have only the relatively centralized public financing mechanism as a source of support. The public sector is simply not designed for such a role. Government must await a politically developed mandate to dispense public funds. In a world with an imaginative and responsible philanthropic sector, a potentially important new idea in a politically sensitive area has many possible sources to turn to for a chance in the marketplace. While the government or several foundations may say "no" to a particular idea, one foundation may support it.

Foundations can also provide a certain continuity of ideas, quite difficult for government. Key policy-making officials change with each administration, or even during a single administration. Foundations have as a special advantage the privilege of being able to look toward the future without rapid turnover in leadership, or at many of the pressing needs of the moment which can deflect even the most stalwart of public officials. Thus foundations can represent a source of support or activity for some long-range or visionary ideas which may add to the quality of life for the generations who follow.

In this chapter we will briefly describe the history of private philanthropy as it emerged in this country, the foundation universe as it exists today in the U.S., and the options for action available to foundations. We will then use the emergence of one foundation devoting its resources to matters of health, the Robert Wood Johnson Foundation, as an illustration of how foundation efforts can contribute to considerations of health policy.

## THE EMERGENCE OF PRIVATE FOUNDATIONS IN THE UNITED STATES

Although philanthropy is as old as civilization, the organized philanthropic institutions known as foundations are a relatively recent development. Most philanthropic activities prior to 1900 had the objective of alleviating the needs of those who were poor and sick, and emerged from religious and local voluntary efforts. Only during this century did the charitable idea of gifts to the needy begin to be expanded by philanthropy which supported the development of new knowledge and understanding which could permanently improve the human condition. Thus today private foundations increasingly support efforts to determine the nature, causes, and ways of removing human misery.[1]

The establishment of large private foundations like the Rockefeller and Carnegie Foundations began only with the start of the 20th century. These institutions had broad flexible and lofty mandates which opened a new era for philanthropy with the compelling challenge to use their resources for imaginative and productive societal purposes.

Private or independent foundations usually have assets derived from the gift of an individual or family, and have as their primary purpose to fund the work of others by making grants. A private foundation is a nongovernmental, nonprofit organization managed by trustees or directors, established to aid social, educational, charitable, religious, or other activities serving the common welfare. They differ from other organizations which also may be called foundations but have purposes other than the awarding of grants, including foundations which administer and operate their own programs, associations for industrial or special groups, and functioning endowments for colleges, churches, or other organizations. They also differ from company-sponsored or corporate foundations which award grants, but derive their funds directly from a profit-making company or corporation.

Private or independent foundations vary widely in the size of their funding resources, the size and number of grant awards, and their staffing. Although all are subject to regulations regarding their management and methods of operation, the proportion of their resources which must be distributed for charitable purposes, and the eligibility criteria which must be met by grantees, they are so heterogeneous that it is not possible to characterize them generically.

Private philanthropic foundations, as well as some other institutions exclusively organized for charitable purposes, are exempted from taxation. No part of their net earnings can inure to the benefit of any private shareholder or individual. They are prohibited from lobbying, carrying on propaganda, or otherwise attempting to influence legislation, or intervening in any political campaign on behalf of any candidate for public office. Furthermore, a private philanthropic foundation must be organized and operate exclusively for religious, charitable, and scientific purposes. These governmental regulations protect the public from use of private philanthropic funds for purposes which are not in the public or national interest.

Within the United States the IRS recognizes 28,000 private foundations. Approximately 21,000 of these foundations are active and made grants in excess of $100 in 1978. Most foundations are relatively small (Figure 1) and confine their grant-making activities within specified geographic areas. Only a few make grant awards nationally or internationally. Only a very few large foundations have professional staff and an even smaller number target on health and medical affairs and have three or more health professional staff (Figure 2).

The programs supported by small foundations have a significant influence on the growth, development or maintenance of community

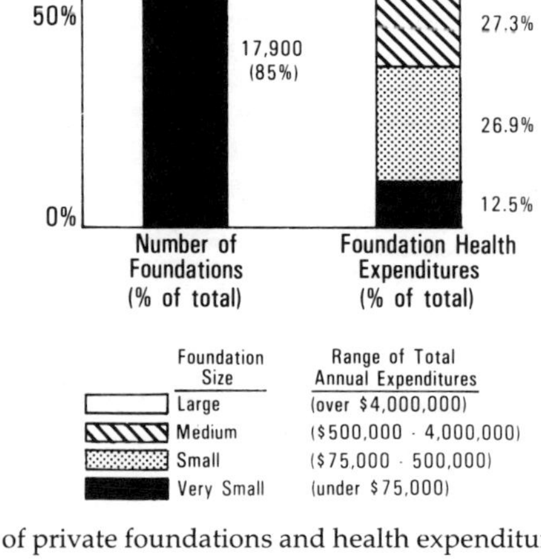


**Figure 1.** Number of private foundations and health expenditures by foundation size.

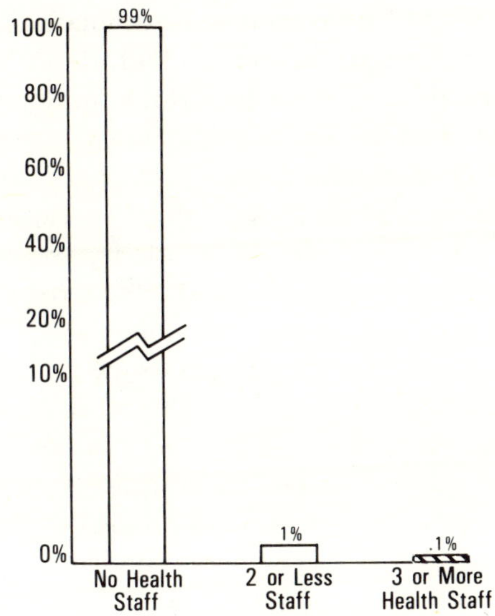


**Figure 2.** Professional health staffing in foundations (full-time).

health activities and institutions. They may stimulate or catalyze larger efforts, but this is not often an objective of these foundations. Larger foundations which are professionally staffed generally have the objective of supporting programs which can advance regional or national efforts.

In the early part of this century philanthropic organizations, particularly foundations, were a major, often the principal, source of funding for new ideas in the health field. In those earlier days funds from philanthropic organizations were responsible for many of the gains in the health sector which are a source of real pride to health professionals and Americans in general. Private philanthropy provided the major funding for medical research leading to the control of hookworm infestation and development of methods for eliminating tuberculosis and poliomyelitis. Philanthropic funds also improved the education of doctors and established public health as a new discipline. Local philanthropic groups developed many of our community hospitals, subsidized medical care for the poor, and spearheaded efforts to improve the health of minority Americans.

Since World War II government support of medical research and health care has steadily risen, and most dramatically during the past 20 years. Thus although private philanthropic foundations have steadily expanded their giving, their funds now represent less than 1% of the total amount the nation contributes to health- and medical-care needs. This change has, understandably, required careful reassessment of the role of philanthropy in this area. Thoughtful people have asked if indeed there remains a role for private foundations in health. But for most, the argument that private risk capital invested in health can make a difference remains persuasive. Public funds must answer the immediate, pressing needs of today. This means providing those funds required for delivering medical services to many Americans—addressing the problems of the here and now. Public funds are also heavily earmarked for support of most of the biomedical research conducted in private and public institutions.

The increasing involvement of government in our nation's health affairs, and the ponderous nature of massive public programs, strongly restricts America's long-term capacity to respond swiftly to changing needs or rapidly developing technologies affecting health care. This makes it difficult for certain kinds of new or untested ideas to gain the support needed for an adequate trial. This also makes it difficult to provide opportunities for individualized, locally tailored initiatives in the health-care area to emerge. In an area as complex and diffuse as health, it seems important to continue to preserve and further pluralistic sources for funding. The public sector alone is not designed to meet some emerging and pressing needs. If health care in the year 2000 is to look different from the health-care system of today, private philanthropic foundations can play an imaginative and important role. Only by having a group of institutions which are free to try routes quite different from those required of government can we be assured that we are protected from the hazards of an increasingly centralized decision-making process.

## OPTIONS FOR FOUNDATION ACTIVITY IN HEALTH

An institution or group having the objective of improving the health of the American people must have instruments or methods to achieve its goals. By no means are all such methods or instruments available to a private foundation. Basically a foundation may:

1. *Support endowment* or the establishment, construction, equipping, or underwriting of the operating expenses of institutions.
2. *Support health-related research.*
3. *Support health professional training and education.*
4. *Support the collection and interpretation of data* to identify programs or new approaches for field testing.
5. *Support demonstrations* by field-testing particular models which look promising.
6. *Support analytic studies* of programs and reports which can make the results of a demonstration or an idea more widely known.

A foundation can support a diversity and variety of program objectives or it can target its funds on just a few problem areas. It can decide to stay with a program for a long or a short time. It will, if it is wise, try to time its programs to coincide with the readiness of society to take action, or encourage efforts to resolve problems so as to generate continuing public or private support. This also means that there should be reason to expect that areas selected for support present significant possibilities for success.

But there are a number of things which foundations cannot do. Foundations are not regulatory agencies. They cannot lobby for particular programs, and they are sharply restricted as to how and to whom they can award grants.

## ONE FOUNDATION'S APPROACH

Although the Robert Wood Johnson Foundation was established in central New Jersey almost 50 years ago as a local philanthropy, it began to develop programmatic goals having national scope only in 1972.

At the outset the decision was made to focus sharply on a limited series of objectives, established after considerable study and wide consultation. The problems selected for these targeted efforts were stated as follows:

1. Improving access to medical-care services for underserved Americans.
2. Improving the quality of health and medical care.
3. Developing mechanisms for objective analyses of public policies in health.

The Foundation also decided to support external evaluations of programs it funded to try to ascertain whether or not they had achieved their objectives.

The evolution of these programs, their successes and failures, have been detailed in a series of publications by both staff and grantees.[2–14]

In 1978 the Foundation staff began a broad review of its work to date and the changes on the American health scene which had taken place during its 9 years of national activity. Those studies and processes whereby the Foundation modified its objectives in 1981 can perhaps illustrate how one foundation formulates, develops, and implements its policies.

## DEVELOPMENT OF THE RWJF PROGRAM FOR THE 1980s

In redefining its mission for the 1980s the Foundation staff undertook an in-depth review of available data to obtain information bearing on several questions.

First, had access to medical care changed since 1972? What individuals or groups were continuing to have the most difficulty in obtaining needed medical care? Second, how had the Foundation service, training, and research program fared? What factors could be identified which might predict success or failure? Third, what would be the political, social, and economic climate in which the Foundation would be working during the 1980s? Fourth, what were the most pressing medical and health problems that would face Americans in the 1980s, and where might the Foundation play a meaningful role in addressing these problems.

Data bearing on the first question was cheerful good news. Over the decade of the 1970s the nation had made impressive strides in improving access to medical care.[15] Those who were poor and minority groups were now seeing doctors as frequently as those who were white and well-to-do. Further, and occurring in parallel with the above, the health of Americans had grown steadily better. Age-adjusted death rates, infant and maternal mortality had fallen quite dramatically during the decade, and 11 of the 15 leading causes of death had shown a decline.[16]

There were, however, certain groups who were lagging behind in obtaining needed services. These included poor children and the poor elderly, and those isolated from the mainstream of American culture by virtue of language, geography, culture, or institutionalization. This led staff to believe we could selectively support efforts to improve access to health care services for such groups.

Study of the political, social, and economic climate in which we would operate in the 1980s led to the identification of a series of trends which we believed might impede further improvements or continued growth in biomedical research, health services research, health professional training, and health and medical services. Some of the forces which might play such a role included:

1. A probable period of slowed economic growth and continuing inflation of uncertain duration.
2. Probable plateauing or actual decline in America's work force productivity, and reduced real personal income.
3. A probable sharp reduction in federal and state health-care expenditures.
4. Continuing rise in health costs leading to increased governmental efforts to control them.
5. Finally, a review of national polls showed a striking shift in public attitudes with a move from concerns about social issues, including health care, to overwhelming concerns about economic issues.

The changes in economic outlook, the shifts in public concerns, and the probable decline in government funds for new developments in health care had important implications for Foundation work. In our judgment these changes would make it very difficult for groups or institutions to initiate new or experimental efforts in the health field to improve access to medical care for those whose needs were still unmet, or to try new ways to deal with shortfalls which continued to exist. Further, academic health centers projected little or no growth in financial support for biomedical research or medical education, and some of these institutions indicated that their financial future might be in jeopardy. The increasing costs of health professional education also made a decrease in the entry of the poor and middle class into medical school seem likely, and the gains in access to health care were at risk unless special efforts were made to preserve them.

This in turn led to some decisions about Foundation strategies which would keep these concerns in the forefront of our activities.

In examining what were the most pressing problems in health which would threaten Americans in the 1980s the problem of chronic disabling disease loomed uppermost.

Although age-adjusted death rates were falling, at the same time there had been a marked increase in the rate of disability affecting Americans.[17] During the same years that the productivity of the American economy was shown to be declining the number of men 45–65 years of age unable to work because of debilitating effects of long-term illness more than doubled. In 1979 disabling conditions caused the loss of 465 million work days—12 times more than the number of days lost from strikes.

Some 80% of health-care resources in the United States are now devoted to health problems caused by chronic disease. Over 60% of patient days of hospitalization in acute hospitals are due to chronic health problems, and over 50% of the visits to physicians are accounted for by chronic conditions.[14] Clearly the problems in health care for the 1980s were to be strikingly related to problems which reduced personal function and resulted in short- or long-term disability. In this sector, technologic and scientific advances are badly needed to prevent or cure those health problems responsible for disability, but there is also an expanding proportion of the population who need medical care to restore or conserve functional effectiveness: to help children learn to succeed in school, to help people continue to work, and to help individuals maintain their independent existence.

Armed with these data the Foundation staff, in consultation with trustees and many experts in the field, reformulated its mission. As it finally evolved, we agreed to encourage and support programs in the following three areas:

1. *Programs to improve access to personal health care for the most underserved population groups.*

   Here the Foundation agreed to target its attention on the 12–15 million Americans who continue to face barriers in getting the medical care they need. A number of these Americans live in geographic isolation or in blighted inner-city areas. Others are culturally isolated because of language or social separateness. Still others are outside the mainstream of medical care because they live in homes for the aged or other institutions.
2. *Programs to make health care arrangements more effective and care more affordable.*

   Here the Foundation agreed to encourage a variety of research, development, and demonstration projects aimed at increasing the capability and efficiency of health services that can bring about tangible improvements in the health status of people and slow the rate of increase in the costs of personal medical care.
3. *Programs to help people maintain or regain maximum attainable function in their everyday lives.*

   Under this rubric the Foundation hopes to encourage approaches which can return people swiftly to independent existence, and help American medicine to move beyond what many regard as its preoccupation with the prevention of death. Here it seems appropriate to concentrate on those kinds of medical- and health-related programs which are targeted on restoring individuals to adequate personal functioning in everyday life.

Because of the limits of the Foundation's resources, a strategy was required to help select from the many, those programs which the Foundation might support. Obviously the Foundation cannot underwrite the ongoing costs of health or medical services, but because in the past, improvements in various sectors of the American enterprise have closely followed significant investments in research and development, the Foundation plans to work in this area.

With these considerations in mind, the Foundation adopted three roles to guide our grant-making activities:

1. The development and testing of new and previously untried approaches to health-care problems.
2. Demonstrations to assess objectively the operational effectiveness and value of selected new health-care arrangements and approaches which have been shown to be effective in more limited settings.
3. Projects designed to promote and broaden diffusion of programs that have been objectively shown to improve health care or make health care more affordable.

## DISCUSSION

How then do a foundation's activities influence the evolution of health policy? As is apparent, the answer is—only indirectly and at the margins. Be that as it may, our work to date leads us to suggest that it plays a role of significant value in our society in the following manner.

First, by identifying and making known a problem area in which it will work, a foundation can help legitimize efforts in a particular area. We believe our focus on the need for more widely available general medical-care services led to more attention being directed toward this area during the 1970s than might otherwise have occurred.

Second, by helping to put demonstrations of new approaches in place, foundations can allow others to examine and test them, thus removing the fear of the unknown or untried.

Third, by conducting careful evaluations of new or experimental programs and reporting the results of those studies, foundations can make those working in a particular sector better aware of what is being done to address a particular problem.

The trick in formulating targeted objectives is to ensure that they are focused on big and important questions, problems, or issues, and that they can encourage innovative approaches to their solution. There are multiple ways in which big questions can be identified and addressed, and there are "many roads to Rome." The objective must be phrased in such a way as to avoid excessive ambiguity so that those who seek support from the Foundation understand its goals. It is also important to ensure that Foundation support is directed to programs which maximize the use of its limited resources. The comparatively small resources of a foundation cannot generally contribute significantly to areas of study receiving very large support from the federal government. Thus a foundation must keep abreast of the program activities of other funding sources so that its own programs are effectively meeting needs or objectives not otherwise being addressed.

A foundation with a well-defined mission and targeted objectives has a privileged opportunity to contribute to national purposes. Hopefully it can help encourage those who develop new knowledge, conduct demonstration programs to allow new ideas to be tested, and disseminate information about new approaches and ideas which appear promising into the mainstream of American life.

## REFERENCES

1. Weaver, W., 1967, *U.S. Philanthropic Foundations: Their History, Structure, Management and Record,* Harper & Row, New York.
2. Rogers, D. E., 1972, One foundation's aspirations, The President's Statement from the Robert Wood Johnson Foundation's 1972 *Annual Report;* Rogers, D. E., 1973, A foundation's continuing education in health affairs, The President's Statement from the Robert Wood Johnson Foundation's 1973 *Annual Report;* Rogers, D. E., 1974, Why a foundation and what does it do?, The President's Statement from the Robert Wood Johnson Foundation's 1974 *Annual Report;* Rogers, D. E., 1975, Reflections on our first four years on the national scene, The President's Statement from the Robert Wood Johnson Foundation's 1975 *Annual Report;* Rogers, D. E., 1976, Progress in American health affairs: Things do change—and sometimes for the better, The President's Statement from the Robert Wood Johnson Foundation's 1976 *Annual Report;* Rogers, D. E., 1977, Where have we been and where are we going?, The President's statement from the Robert Wood Johnson Foundation's 1977 *Annual Report.*
3. Rogers, D. E., 1978, *American Medicine: Challenge for the 1980s,* Ballinger, Cambridge.

4. Andersen, R., and Aday, L. A., 1978, Access to medical care in the U.S.: Realized and potential, *Med. Care* **16:**533–546.
5. Delbanco, T. L., and Parker, J. N., 1978, Primary care at a teaching hospital: History, problems, and prospects, *Mount Sinai J. Med.* **45:**628–645.
6. McDermott, W., 1978, Health impact of the physician, *Am. J. Med.* **65:**569–571.
7. McDermott, W., 1978, The public good and one's own, *Perspect. Biol. Med.* **21:**167–187.
8. Rogers, D. E., and Blendon, R. J., 1978, The academic medical center: A stressed American institution, *N. Engl. J. Med.* **298:**940–954.
9. Aiken, L. H., *et al.*, 1979, The contribution of specialists to the delivery of primary care, *N. Engl. J. Med.* **300:**1363–1370.
10. Moloney, T. W., and Rogers, D. E., 1979, Medical technology—a different view of the contentious debate over costs, *N. Engl. J. Med.* **301:**1413–1419.
11. Aiken, L. H. (ed.), 1981, *Health Policy and Nursing Policy,* McGraw-Hill, New York.
12. Aiken, L. H., *et al.*, 1980, Evaluating a foundation's health program, *Eval. Prog. Planning* **3:**119–129.
13. Blendon, R. J., 1978, The prospects for state and local governments playing a broader role in health care in the 1980s, *Am. J. Public Health* **71:**9–14.
14. Cluff, L. E., 1981, Chronic disease, function and the quality of care, *J. Chronic Dis.* **34:**299–304.
15. Aday, L. A., Andersen, R., and Fleming, G. V., 1980, *Health in the U.S.: Equitable for Whom?,* Sage Publications, Beverly Hills.
16. McDermott, W., 1981, Absence of indicators of the influence of physicians on a society's health, *Am. J. Med.* **70:**833–843.
17. Cluff, L. E., 1981, Conserving function: The goal of medical care, Jacob Yules Visiting Professor Lecture, Tufts University School of Medicine, May 1981.

# 6

# Differing Approaches to Biomedical Research

## The NIH, the Academic Medical Center, and the Pharmaceutical Industry

THEODORE COOPER and SUSAN BENNETT

In order to study different approaches to research, one might visit scientists working in an intramural laboratory at the National Institutes of Health (NIH), scientists of similar kind in an academic medical center, and others of the same discipline in a laboratory of a pharmaceutical firm. One would initially find little to write about. The physical settings and the instrumentation inside the laboratories would be similar. The kinds of people milling about, their jargon, the smells, and the sounds would be indistinguishable in going from one laboratory to another. If, however, the visits allowed one enough time to see how each of the scientists spends his or her time over the course of several days, one would begin to see important differences, differences reflecting the missions and philosophies of the respective parent institutions as well as the missions and philosophies of the financial supporters of the research being undertaken by each of the scientists.

It would seem logical to assume that the scientist in each setting has chosen to work in that particular setting because he or she shares the basic philosophy of the parent institution and the commitment to its goals. To some extent that assumption would be true. But today the orientation of the parent institution appears less important than

THEODORE COOPER and SUSAN BENNETT • The Upjohn Company, Kalamazoo, Michigan 49001.

the economic determinants of a research laboratory's existence. In other words, good scientists want to go where they can do good science with some assurance of economic stability. Of course this would not apply to everyone who calls himself or herself a scientist. There are many persons now working in laboratories or agencies who are involved in activities other than investigation. For example, some people currently managing drug-development programs in pharmaceutical firms who had been bench scientists are still called scientists. There are people at the NIH administering grant and contract programs who still prefer to be classified as scientists because they were once laboratory scientists. And there are individuals administering research programs in academic medical centers who do no investigation themselves but who nevertheless prefer to be known as scientists.

Not too long ago one could see a good deal of difference with a brief visit to an actual laboratory in the three settings. The operations were smaller and reflected more directly the character of the parent institutions. Basic research was a less obvious thrust of the industrial laboratory or the clinical department of a medical school. In fact the designation "academic medical center" is of relatively recent origin. Even at the NIH laboratories, disease-oriented labels dominated the apparent thrust of programs. As a result one immediately understood that the laboratory of the pharmaceutical firm was concerned with the yield of a marketable product; the laboratory of the clinical department in a medical center was studying a patient, a therapy, or a disease; the basic science department was studying chemistry and the structure and functions of various levels of biological systems; and the NIH was akin to a mixture of the latter two.

Today all institutions that depend on modern biotechnology are aware that their intellectual vigor will depend on being able to enfold into their programs rapidly evolving "new" knowledge concerning molecular and cellular systems, systems integration, organ structure, chemical transformation, and so on. In other words, the ability to develop or apply new techniques or treatments is greatly facilitated by strong foundations in basic research. Therefore one sees similar laboratory settings and activities occupying greater portions of laboratory operations in industry, academia, and the government.

It is becoming increasingly more expensive to stay on what is known as "the cutting edge" of basic research, of discovery. The processes by which new knowledge is developed transcend institutional, geographical, regional, and political boundaries. The analytical tools are large and complex. The supporting environments require definition and control to extents not previously needed. The technical assistance involved presupposes greater educational backgrounds of the personnel. More personnel support a single project. Experimental animals require special breeding, handling, and maintenance. Laboratory buildings need more sophisticated engineering. Consequently the academic medical center has to limit its aspiration. Few centers can choose to have first-rate programs in every area that might interest the faculty and staff. The scientist at the NIH must also choose between areas of interest because federal appropriations do not grow at rates sufficient to allow the exploration of all areas. The industrial laboratory also must pick and choose among opportunities. Few firms can afford not to set priorities because of the costs of research and development.

Each of the three institutions addresses these challenges differently. Herein lies the difference in the approaches to biomedical research. An academic medical center of the first rank, a first-rate laboratory in industry, and the splendid laboratories at the NIH have a good deal in common, but there are also important differences. Each has a different mission to fulfill or goal to achieve; each operates from a different base. The mix of activities of the various programs in each of the three settings, and the mix of schedules of the scientists in each, reflect the difference in approach of the respective organizations.

## THE NATIONAL INSTITUTES OF HEALTH

The broad mission of the NIH is to "promote the health and well-being of . . . citizens" by encouraging and supporting research that will lead to biomedical innovations. Financed by a direct appropriation of public funds by Congress, government-sponsored biomedical research is a vital undertaking that encompasses a broad range of social forces. The government is sometimes in the best position to pull together the best sources of science, free of vested interest and relatively free of political bias.

The growth of innovation—in the form of new and/or improved products and technologies—surged during World War II, when a true partnership existed among the government, industry, and academia, designed to satisfy the nation's pharmaceutical needs as well as the specific urgent medical requirements of the war effort. Cooperative research efforts flourished and the government adopted a supportive, positive attitude toward industrial research. A predictable result of this spirit of accommodation and collaboration for the common good was a remarkable shortening of the span of time required for a new discovery to be translated into a useful product or technology.

One of the largest and most respected medical research centers in the world, the NIH began almost 100 years ago as a one-room laboratory. In addition to operating an intramural program through its 11 separate institutes and the National Library of Medicine, all located on the NIH campus in Bethesda, Maryland, the NIH administers funding to a variety of research activities in academia and private institutions.[1]

The 11 institutes of the NIH were established over a span of some 26 years, as follows:

| | |
|---|---|
| 1938 | National Cancer Institute |
| 1948 | National Heart, Lung, and Blood Institute |
| 1948 | National Institute of Dental Research |
| 1950 | National Institute of Arthritis, Metabolism, and Digestive Diseases |
| 1950 | National Institute of Neurological and Communicative Disorders and Stroke |
| 1955 | National Institute of Allergy and Infectious Diseases |
| 1962 | National Institute of Child Health and Human Development |
| 1962 | National Institute of General Medical Sciences |
| 1968 | National Eye Institute |
| 1969 | National Institute of Environmental Health Sciences |
| 1974 | National Institute on Aging |

Not an institute, but an integral part of the NIH nonetheless, the

National Library of Medicine was founded in 1956. And since it is not limited entirely to supporting U.S.-based research, the NIH has offices in Europe, Asia, and South America.

The NIH reached its peak growth in the early 1960s and has pretty much leveled off since then, with "blips" of growth reflecting a national interest in specific diseases or health issues, e.g., infectious disorders and the aging process. Future program developments will continue to be responsive to public concerns.

To facilitate both the program planning and the budgeting processes, NIH activities are structured around four major classifications:

1. *Science Base.* All efforts that comprise the search for new knowledge; discovery. Research grants/contracts and some center-based and intramural projects would fall under this designation.

2. *Clinical Application.* Developmental, assessment, and verification of potentially useful innovations. The largest activity in this classification is the program of clinical trials that are necessary to test the safety and efficacy of medicinals and other therapeutic interventions intended for human use.

3. *Transfer.* Dissemination of new knowledge. The function of this realm of activity is to get new knowledge into the hands of professionals who can put it to practical use.

4. *Training.* Locating, encouraging, and training people for careers in biomedical research. The future of biomedical science is dependent on adequate numbers of well-trained, motivated investigators.

## THE ACADEMIC MEDICAL CENTER

The mission of the academic medical center is to provide a training environment for the doctors and scientists who engage in the research/discovery/development process. It is a scholarly endeavor that reflects the basic perception of academia to discover new knowledge and to transmit it to others.

Academic scientists work in collaboration with the pharmaceutical industry to perform vital work in the link from basic research to the translation of that research into useful therapeutic entities. The laboratory in an academic medical center is usually funded from a combination of sources: the parent institution; federal agency extramural programs; private industry; professional fees; state, local, and federal taxes; and not-for-profit institutions.

The training environment is organized to carry out the roles of teaching, research, and patient care—functions for which the academic medical center is uniquely suited. The goals of the NIH, academia, and industry all come together in academic medical centers.

One of the most important ongoing needs in academic medicine is the recruitment and training of manpower. With available resources essentially in a no-growth pattern over the past several years, it is absolutely essential that "academic" institutions like the government and industry identify ways to stretch their funds. Manpower salaries comprise a significant portion of research budgets. Since we cannot imagine that reducing salaries would be greeted with any degree of acceptance, the first alternative that comes to mind is a reduction in the number of clinical investigator recruits. That of course would probably require that the number of institutions engaged in clinical investigation would also be reduced. In turn that would require that the personnel recruited would have to be those with exceptional talent who could be trained to broad competence. To accomplish all of this will demand excellence in planning and execution. This is only one possible alternative; it may not be the best one. But it will have been useful if it stimulates the development of others.

## THE PHARMACEUTICAL INDUSTRY

The ultimate goal of the pharmaceutical industry is to maximize the investment of its stockholders, either by performing independent research that will lead to useful, profitable products and technologies, or by developing the research discoveries of government or academia into useful, profitable products and technologies. The laboratory of the pharmaceutical firm depends on profits for its existence and operates primarily from a base of private funds, generally as a direct expense of the firm (although occasionally there is some funding from a federal agency through a contractual agreement).

The pattern of growth in research, discovery, and innovation that was enjoyed by the pharmaceutical industry in World War II continued into the 1960s. Since 1970, however, research and development as a percentage of sales has declined in a number of pharmaceutical companies. In 10 companies for which figures are available, average research and development expense, expressed as a percentage of sales, was 7.6% in 1970.[2] The comparable 1979 figure for those companies is 6.7%, a decline of 0.9%. A number of reasons are given for the decline, including a lower rate of return on research and development costs, which have risen dramatically while new product introductions have slowed; increased regulatory restrictions and guidelines; and increased diversification into nonpharmaceutical activities.

In reality the monies spent for research and development by the pharmaceutical industry could be viewed as capital investments on which a return is not realized until a new product or technology is introduced. In practice, however, they are treated as operating expenses, thus producing a negative effect on earnings.

For these reasons and others, it is understandable that pharmaceutical companies may question the wisdom of relying on the discovery and development of new compounds in their own laboratories for a major portion of their future income. There *are* some companies which do embrace that philosophy, but we suspect that in the years ahead their number will shrink and the burden of biomedical research will fall to fewer pharmaceutical companies, with substantial pressure for greater involvement of the government and, through the government, of academia.

## DISCUSSION

Biomedical research is an investment in the human condition, and it pays huge dividends. The products and technologies that result from biomedical research save lives that otherwise would have been lost and improve the quality of life for countless numbers of human beings. Ordinary cost–benefit analysis procedures are inadequate to account for the advances in health science and technology that are attributable to the efforts of biomedical research. How does one assign a reasonable "cost" to a life? The philosophical, moral, and ethical implications inherent in attempting to arrive at an answer to that question are enormous. What we can find comfort in, though, is that *whatever* the cost of biomedical research, the price extracted for *not* funding it would be much greater. In the final analysis, effective medicinals and therapeutic interventions are the most efficient and least costly forms of personal health services.

The essential ingredient of a fine laboratory is quality personnel. The fact that we now have good laboratories throughout the various sectors of the biomedical system of the United States can be attributed largely to the success of the NIH. Both the intramural and extramural programs of the NIH produce excellent scientists as a primary objective. Of course many of these scientists were educated and trained in hundreds of different academic settings other than the intramural laboratories of the NIH. The NIH and the academic medical center have an important goal in common: the making of scientists. That is not to say that a good scientist cannot be trained in the laboratory of a pharmaceutical firm. Obviously some of these laboratories have trained and are training excellent new people. However, the industrial laboratory does not have the production of new scientists as a *primary* objective.

In having carried the responsibility for the making of scientists, the NIH and the academic medical centers are charged with controlling the number of scientists. It has been on such an issue that the relationships between the government and academia have become strained. Academic institutions have as a continuing goal the education and training of investigators, whereas the government agencies no longer view the production of scientific personnel as a priority. As the government has reduced its funding to academia, the academic units must face the need to reconsider their plans and approaches. The availability of personnel is important to the pharmaceutical industry. While the firms carry out in-service development and training programs, they seek their talent from academic medical centers, from nonmedical parts of the university complex, and from the government. Recognizing the impact of the reduction in federal support for research training, the pharmaceutical firms, collectively and individually, have committed resources to fellowship activities. In some instances these opportunities are even taking the form of postdoctoral programs in pharmaceutical laboratories. The industry hopes to interest more talent and more interest by demonstrating that the core scientific programs of the technology base of a firm are on a par with the programs of academic medical centers and the government. Industry realizes that formerly there was a stigma attached to working in the laboratories of the pharmaceutical industry, a stigma not present in other parts of the industrial sector. For example, Bell Laboratories has for years been viewed as outstanding.

It is also of some interest that a financial motivation is becoming less a handicap to a distinguished scientific career. Recognizing this changing pattern of financing, a variety of activities related to biomedical research will be important to structuring stability and quality into research programs over the next two or three decades. Scientists are discussing targets, both scientific and financial, with greater patience and with considerably more understanding; thus there is better direct conversation between scientists of academia and those of the pharmaceutical industry.

The largest ongoing cost of any research organization is personnel. On the surface, people tend to think that scientists spend all their working time doing laboratory work, thinking, and writing. Therefore it is often concluded that the personnel budget for research reflects all the time and effort being spent by scientists solving problems. Not so. The expansion of a research budget has to a significant extent reflected the socialization of the scientist. This process occurs in academic medical centers to the largest extent, followed by industry and then by the intramural laboratories of the NIH. The socialization process is a manifestation of the philosophy that organizations will thrive best if the people are participating in their own governance. Consequently in all three institutions there have developed a constellation of committees which involve more and more of the scientists' time and energies. It is remarkable how similar such committees seem in the different settings. They talk about familiar things in the same way over long periods of time. The scientists complain weakly how their time is being wasted, but they go back for more with stunning regularity.

One result of this process, and it is not the intended result, is that more people are hired to keep up the output of the laboratory. Managers rarely incorporate this feature into their analyses of personnel needs. Instead they focus on opportunity, composition, and working conditions. The managers also attempt to strike a cultural note. They attempt to normalize the number of technical assistants available for each scientist. It is in vogue to minimize the "scut work" of a scientist, sometimes to the extreme that the scientist has to involve himself or herself minimally in the actual execution of laboratory experiments. The result is more expense.

We emphasize the expense feature of the approaches to research because there is an increasing leverage put on the directions of biomedical research as a result of budgetary pressures. Furthermore, it has often been argued that federally sponsored research in academic settings is much more cost effective than industry-sponsored research. The questions of measuring benefit aside, it is important to look at what constitutes the costs. Until recently another feature of personnel density distinguished the industrial laboratory from the academic and the NIH laboratories: the need to comply with certain regulations that were developed to ensure the quality of the data presented by a pharmaceutical firm to the Food and Drug Administration.

Good Laboratory Practices regulations (commonly referred to as the GLP) have resulted in substantial upgrading of facilities and personnel—their numbers, their level, and their practices. It was only a matter of time before the approach spilled over into the academic sector. Even the NIH must look to its practices, since some of its data find their way into regulatory decisions. Consequently larger numbers of people in research are less directly involved in discovery. More are now "support personnel." The entire research setting, whether it be an academic medical center, industry, or the NIH, is therefore in a cultural change—an expensive change that the academic medical center is having difficulty financing and one which, because of the inability to maintain an appropriate capital program, has resulted in the NIH no longer being a showpiece collection of laboratories. The technology-based firms of the pharmaceutical industry on the other hand are going forward with substantial new commitments to capital projects to finance their research programs and concomitant planned expansions for personnel and programmatic activities. In association with these commitments have been significant changes in appreciation on the part of senior management for the importance of basic research and of attracting talent to their technical areas. The new interest is not born of altruism, nor is it an attempt to provide employment for scientists who are unable to locate secure, well-supported positions in academia. It is a recognition that the ability to compete in the marketplace will depend on success in the laboratory.

Many large firms have developed their product lines by acquisition and licensing from small firms or academic laboratories. Such firms were able to develop a chemical or biological entity because of the availability of funds and the type of staff and programs required to satisfy the requirements for product registration. While opportunities to acquire and license new entities will still exist and will probably multiply, the ability to evaluate, modify, and develop the emerging new groups of compounds will require new skills. Fewer scientists are likely to want to sell their ideas to firms which have little in-house capability to contribute to the development of products.

It is also recognized that developing new products from discoveries made within the laboratories of the firm will afford advantages. The advantages involve patents, market potential, and competitive position, as well as a distinct financial advantage. Obviously one can retain more profit if one is not required to pay royalties. The impact of this change in awareness then is a new comprehension within firms of the need by the marketing, manufacturing, legal, and financial groups to know more about new research activities. Some education of these groups is achieved by attending national and regional seminars. However, most of the process of education is being accomplished by in-house interaction among the researchers. There is a different relationship between business and science, resulting in an increased awareness of the importance of research. It is opposite to what many have experienced in the settings of academia and government. The concern has been that the attitude of the patrons of academia and government programs has been skewed toward skepticism and estrangement rather than toward enthusiastic appreciation of the potential of biomedical and behavioral sciences in the near future.

If you follow scientists to their committee meetings, you will find that these sessions deal with many concerns. There are many in common, as we have said: animal care, human experimentation procedures, promotion, seminar selection, equal employment opportunity, awards, and so on. The differences reflect not only different missions, but also different problems or concerns. For example, in the academic medical center, particulary "private" (i.e., non-state-supported) ones, hours are spent on devising schemes for institutional fundraising. The scientist in a pharmaceutical firm does not do that. The scientist at the NIH might spend hours preparing materials for presentation to Congress. Groups in the pharmaceutical firm are also evaluating the potential of a given research finding for formulation into a product. While scientists in any setting might devote some time to the practical implications of their findings, until recently the financial potential of the observations was of relatively little interest.

Therein lies the fundamental difference in approach. In assessing financial potential or in evaluating what therapeutic application can emerge from a finding, many pedestrian things must be done in the laboratory. In order to attempt to minimize the burden of routine, some groups attempt to allocate time for discovery. This practice is rarely satisfactory. Product programs have found ways to attract individuals who are well suited to development activities. These individuals have often been productive laboratory scientists who seek a career change or who are interested in bringing an idea to completion.

Accolades are awarded for the discovery of something "new" in any setting. In the pharmaceutical firm there are also accolades for bringing the idea to a commercial success. There is an additional dimension of incentive. Yet all the activities are often called research and there emerges the notion that value is only placed on development of the product for the marketplace.

The other facet of commercialization is reflected in the areas selected for research. In academia the administration may consciously choose a person to head a department because they see a need for having an area taught or studied. For example, more medical centers are developing departments of genetics because the administration recognizes the need for this discipline in a modern medical academic complex. In other instances programs evolve because of the availability of resources or because of the preference of search committees for individuals sought on the basis of reputation, quality, or potential. In industrial settings individuals are selected to provide activity in a given area identified in the company's strategy for the future. It is common nowadays for companies to seek out molecular biologists, geneticists, immunologists, and biologists. The reasons are obvious. These corporations will be attempting to develop products that will derive from such research. Or a company may decide to emphasize products for a specific disease and will therefore recruit people with experience and interest in that area. An example is the interest some firms have expressed in cancer chemotherapeutics. It would be expected in this latter situation that the research of an individual recruited for a cancer program would be targeted to studies related to cancer. If a firm chooses to exclude itself from a given spectrum of products, scientists within laboratories will not be encouraged to pursue studies in the excluded area.

The laboratory of a pharmaceutical firm is expected to produce marketable products. But it is recognized by management that a good research program needs to produce good science as the raw material from which those marketable products are developed. Academic and government laboratories are not usually focused on a product. However, some "targeted" or "contract" government programs have been product oriented and have been pursued in collaboration with academic laboratories. Such programs at the NIH have been criticized from time to time because of their orientation, method of administration, evaluation, or quality of work. While in some circumstances the criticisms have been warranted, more often than not the concern has stemmed from the availability of funds. The academic community puts a higher value on investigator-initiated research, partly in self-interest, but also partly because they believe that the research supported by the investigator-initiated grant is of higher quality. There has also been the feeling that more quality research is obtained for a lesser price.

The academic community has little basis on which to compare

the yield per dollar from an NIH grant with the yield per dollar from a direct allocation to a scientist in an intramural laboratory at the NIH. It has even less experience or basis on which to evaluate "the bang per buck" that is allocated to an industrial laboratory. There is some experience by academicians in attempting to assess what a company will pay for research in the academic laboratories. It is likely that a corporate project *budget* proposal will be scrutinized much more extensively than will a similar grant request to the government. Nevertheless it is apparent that industry is quite willing to pay what is necessary to get the help and collaboration it seeks from scientists in the academic medical center. It is also likely that there will emerge broader agreements under which industry will support research in the laboratories of universities, particularly those in academic medical centers.

We can see from Table I that the federal government is easily the largest purchaser of health research and development. The major portion of federal monies are administered by the NIH to its network of intramural programs in Bethesda and to a variety of academic endeavors. But industry is also financing enormous amounts of health-related research. This industry-sponsored research is not spread evenly through the pharmaceutical industry, but rather is concentrated among a relatively few companies that remain committed to the philosophy that their future depends on their independent capacity to discover and develop new or improved compounds.

The amounts in Table I are shown as current dollars; when adjusted for inflation we can see that little or no growth is occurring, that the rate of inflation is about equal to the increased amounts available for research activities.

Numerous reports and explanations of the current political and financial problems of the NIH have appeared in the literature. The major emphasis remains a commitment to the support of basic research as a priority. It is also apparent that attempts are being made to stabilize the national program agreements to limit program size, but not cost. Such maneuvers, in our opinion, are likely to fail in their objectives. However, any attempts to gain reassurance of federal support of biomedical research are welcome.

**Table I.** U.S. Funding for Health Research and Development, 1969–1979

| | Source of funds | | | | | | | |
|---|---|---|---|---|---|---|---|---|
| | Federal | | Industry | | Other[b] | | Total | |
| Year | Amount[a] | Percent change over prior year | Amount[a] | Percent change over prior year | Amount[a] | Percent change over prior year | Amount[a] | Percent change over prior year |
| 1969 | $1674 | — | $ 754 | — | $357 | — | $2785 | — |
| 1970 | 1667 | −0.4 | 795 | 5.4 | 384 | 7.6 | 2846 | 2.2 |
| 1971 | 1877 | 12.6 | 860 | 8.2 | 430 | 12.0 | 3167 | 11.3 |
| 1972 | 2147 | 14.4 | 925 | 7.6 | 455 | 5.8 | 3527 | 11.4 |
| 1973 | 2225 | 3.6 | 1033 | 11.7 | 477 | 4.8 | 3735 | 5.9 |
| 1974 | 2754 | 23.8 | 1171 | 13.4 | 506 | 6.1 | 4431 | 18.6 |
| 1975 | 2833 | 2.9 | 1306 | 11.5 | 550 | 8.7 | 4689 | 5.8 |
| 1976 | 3068 | 8.3 | 1439 | 10.2 | 579 | 5.3 | 5086 | 8.5 |
| 1977 | 3396 | 10.7 | 1592 | 10.6 | 611 | 5.5 | 5599 | 10.1 |
| 1978 | 3811 | 12.2 | 1775 | 11.5 | 668 | 9.5 | 6254 | 11.7 |
| 1979 | 4321 | 13.4 | 2026 | 14.1 | 700 | 4.8 | 7047 | 12.7 |

[a]Dollar amounts in millions.
[b]State and local funds, nonprofit organizations.

Analyses of sector support and program size from various agencies are subject to a variety of interpretations. While failing to stay current with inflation in the general economy, the size of the NIH appropriation has continued to increase. The allocation of resources to the support of research in the private sector has also increased in absolute dollar amounts, whereas the percentage of total health expenditures in the country dedicated to research has fallen.

Within this statistical morass is the clear indication of a change of approach in all three settings: the NIH, the pharmaceutical industry, and the academic medical center. Knowledge is expanding, commercial opportunities are increasing, attitudes of scientists are changing, and the stability of academic and federal programs is diminishing. When viewed together with the influence that technology itself is having on the approaches to research, we conclude that new opportunities for interaction and collaboration among scientists working in the laboratories of academia, industry, and government will emerge and that the new arrangements will prove productive.

## CHALLENGES

Changes in patent-assignment policies could shorten the length of time required to move a compound from discovery to translation into a useful therapeutic entity and might increase the productivity of academia–industry collaboration. At present, patent policies at most academic institutions conflict with those of the NIH. The university scientist-inventor generally is required to assign patent rights to the university; the university then handles the licensing of patent rights according to its individual policy. The NIH patent position in the past has been generally interpreted to mean that patent exclusivity must not be assigned to a specific company. The drawback here is that it is difficult to interest a pharmaceutical company in investing in the development of a compound unless the company is assured of having exclusive patent-licensing rights. There are indications that the NIH is changing its position on patent licensing, but that position must be clarified if industry is to be encouraged to enter into compound development at an early stage, when large amounts of funding and a commitment of project time are required.

One industry spokesman offers the intriguing suggestion that changes in existing patent policies might offer the additional benefit to universities of infusing significant amounts of income from patent-licensing activities into their research programs.[3] It is tantalizing to think that such income might in time significantly free our academic institutions from overwhelming dependence on the government for the support of university-based research.

The current patent structure in the U.S., whereby patents are applied for and granted when a chemical entity is shown to be potentially useful, poses a major problem to the pharmaceutical industry. By the time a compound completes the extraordinarily complex and regulated development process (and only a small percentage of them do) and is introduced on the market, the patent life is severely reduced, giving the company a relatively short time to recover the research and development costs.

New drug development is lengthy, costly, and complex. This is due in part to the government's regulatory constraints. Government-specified study methodologies, informed consent requirements, patient population restrictions, and institutional review boards (IRBs) are specific examples of government regulations that sometimes contribute to problems in the new drug development process.[4]

No one would argue that safeguards must be provided to protect the public interest, but if those safeguards are excessive the whole drug development process may become jeopardized. In the matter of government-specified study methodologies, for example, insistence on certain types of complex studies supported by vast quantities of case records is sometimes viewed by investigators as being unnecessarily burdensome and stifling. The original intent of the requirement for adequate, well-controlled investigations was to prove the safety and efficacy of a drug when used as intended. Informed consent is vital to first-rate clinical research, but we should question the necessity for complicated, inordinately time-consuming informed consent procedures. By the same token, limitations on clinical research involving fetuses, children, pregnant women (and those women who have a high potential for becoming pregnant), and the mentally impaired are prudent; exclusion of prison populations from clinical research on the presumption that prisoners are incapable of voluntary informed consent may be of questionable validity. IRBs, which supervise pharmaceutical-industry-sponsored institutional research activities, are struggling under heavier work loads and increased supervisory accountability and legal responsibilities. This is resulting in expensive and time-consuming delays in approving institutional research activities. There is little documentation that such requirements have improved the quality of clinical research.

The pressures on governmental decision-making emanate from a variety of sources—from Congress, from the public, from special-interest lobbies, from the administration, from academia—and there are indications that the federal government is under substantial pressure to assume increasing responsibility for the support of biomedical research. Less available monies for university funding and an increasingly complex regulatory environment that dominates the pharmaceutical industry are seriously constricting research activities in these arenas. As a result the federal government is being asked to provide assistance—assistance in the form of additional funds for university-performed research and assistance in the form of regulatory revisions, tax relief, and increased patent protection for the pharmaceutical industry.

There is a potential for conflict of interest whenever funds are provided for biomedical research. That is true whether the funds are derived from industry, from academia, or from the government. An investigator is always faced with the question of where his or her ultimate loyalty rests. Recognizing that the potential for conflict of interest exists, we must take on faith the personal integrity of the investigator to a commitment to good research. If we cannot do that, then biomedical research is forever suspect.

The increasing demands for more studies of more kinds in man (rather than in experimental animals) will bring the academic medical center and the pharmaceutical industry closer together. The scope, diversity, and number of subjects needed for studies will increase; therefore the need for creative collaboration between pharmaceutical laboratories and medical centers will also increase. And the nature of these arrangements will change. Currently studies are specified on a project basis. It is likely that broader arrangements will emerge and that fast-turnaround concepts will be sought. Arrangements which support ongoing availability of services will emerge. Academic programs will seek means to structure such arrangements in order to attempt to stabilize their operations in the near term. Objectives for the longer term are less clear. The more technologically based institutions are exploring options to diversify their financial bases of operation, but some academic medical centers cannot think beyond the notion of federal support as the only basis on which to predicate their future.

It is important to consider what these changing approaches will mean to information transfer. Will there be more secrecy in research fostered by different financial incentives or arrangements? Some observers feel that the greater involvement of scientists in academic medical centers in profit-making ventures is resulting in less open communication. Reports are becoming less informative in the more competitive fields. Academic scientists who collaborate with industry have always recognized the possibility that some of their industry-sponsored work might not reach the open literature in the same time frame or in the same form that they might have employed in reporting results of research sponsored by the NIH or by the university. Today even scientists who are not sponsored by industry are evaluating the potential commercial value before publishing their research findings. It is difficult to ascertain what pattern will emerge from the current instability and change. It is certain, however, that one outcome will be an awareness of greater self-interest in the scientist, irrespective of the setting of his or her laboratory.

There are no easy answers to the problems facing the biomedical research community. Additional federal funding could be provided—but at the probable expense of other worthwhile programs. Regulations, tax policies, and patent laws could be revised—but the utmost caution would have to be exercised to ensure that the trade-offs implicit in the revisions were in the best public interest. If we are truly committed to a program of biomedical research, however, it behooves all of us—the federal government, academia, and the pharmaceutical industry—to consider how we can best solve the problems confronting us without necessarily looking to the other fellow to bail us out. Planning, ingenuity, imagination, and determination may be our most valuable resources.

## CONCLUSION

It is imperative to retain the differences in these approaches to developing science for the health system. It is equally imperative to emphasize and encourage the ability of the government, academia, and the pharmaceutical industry to work together. Although each approach makes vital contributions, we should constantly question whether each can do more, or better, or differently. Our expectations are enormous; the problems and needs of the poor, the elderly, the hungry, the emotionally and physically impaired are many. As the American standard of living has risen, so has our perception of what is acceptable in the realm of health also risen.

We have no reason to believe that the days of routinely increased research budgets will return. Instead we will have to rely on imaginative and innovative thinking that will lead to improved productivity in order to stretch the research funds that are available. The pressures to contain costs and to justify expenditures for manpower, facilities, and equipment will have to be accommodated. Perhaps the number of funded projects will decrease; perhaps individual projects will not be funded as well as we would wish. That does not mean, however, that the *quality* of research must decline. What it *does* mean is that we must recognize that funds for research budgets indeed are finite. All of us who engage in biomedical research will have to accept the responsibility for identifying and effecting economies without compromising excellence in our research efforts.

## BIBLIOGRAPHY

Fredrickson, D. S., 1981, Biomedical research in the 1980's, *N. Engl. J. Med.* **304**(9):509–517.

Goldberg, L. I., 1980, An academic investigator's view of collaboration with industry, *Perspect. Biol. Med.* **23**(2, part 2):S89–S91.

Helms, R. B. (ed.), 1981, *Drugs and Health: Economic Issues and Policy Objectives*, American Enterprise Institute for Public Policy Research, Washington and London.

Industry and the universities: Developing cooperative research relationships in the national interest. Report of the National Commission on Research to the 42nd Meeting of the Advisory Committee to the Director, NIH, March 16–17, 1981, Bethesda, Maryland.

Kass, E. H., 1980, Factors affecting research support and career decisions in clinical investigation, *Perspect. Biol. Med.* **23**(2, part 2):S44–S52.

Marks, P. A., 1980, Comments on federal policy in biomedical research, *Ann. N.Y. Acad. Sci.* **335**:176–177.

Muller, S., 1981, Resources for research medicine, *Science* **212**(4500):1219.

Schneyer, S., 1981, You can plan to look for gold, but you can't plan to find it: Comprehensive planning for health research revisited, *Clin. Res.* **29**(2):95–100.

Stetten, D., Jr., 1980, Research in a no-growth economy—can excellence be preserved?, *Perspect. Biol. Med.* **23**(3):357–369.

Swisher, S. N., 1980, Manpower training for fundamental clinical research, *Perspect. Biol. Med.* **23**(2, part 2):S62–S78.

## REFERENCES

1. Fredrickson, D. S., 1978, The National Institutes of Health: Yesterday, today, and tomorrow, *Public Health Rep.* **93**(6):642–647.
2. Grabowski, H. G., and Vernon, J., 1981, The determinants of research and development expenditures in the pharmaceutical industry, *in: Drugs and Health: Economic Issues and Policy Objectives* (R. B. Helms, ed.), American Enterprise Institute for Public Policy Research, Washington and London, pp. 3–20.
3. Undenfriend, S., 1980, Improving the scientific base of clinical research by facilitating collaboration among academia, the pharmaceutical industry, and government, *Perspect. Biol. Med.* **23**(2, part 2):S83–S88.
4. Nelson, A. J., 1980, Regulator encumbrances to pharmaceutical research, *Perspect. Biol. Med.* **23**(2, part 2):S150–S153.

# 7

# World Endemic Disease

## Costs and Potential Fiscal Benefits of Medical Research

GRAHAM V. BROWN and G. J. V. NOSSAL

Parasitic diseases are rampant throughout the world. In 1976, according to WHO estimates, 1.2 billion people lived in areas where malaria continued to flourish despite some programs for control in areas where three-quarters of these people lived.[1] An estimated 180–250 million people are infected with schistosomiasis, and in 1971 about 650 million people were thought to have ascariasis (roundworms).[2] Some campaigns to control malaria were partially successful in the 1950s and 1960s but there has been a resurgence of the disease in many countries including Indonesia, Sri Lanka, and the Indian subcontinent. The number of reported malaria cases increased by over 230% between 1972 and 1976,[3] and there is the ever-present threat of epidemic disease in a nonimmune population.

Infection, poverty, and malnutrition dominate the health problems of poor countries, and some would argue that the process of social and economic development will inevitably lead to a decline in the morbidity and mortality associated with these diseases and that specific input from the health sector will provide little further benefit. In Europe, a major part of the decline in the morbidity from infectious diseases (e.g., pertussis, tuberculosis, and scarlet fever) occurred prior to the development of specific measures for prevention or treat-

GRAHAM V. BROWN and G. J. V. NOSSAL • The Walter and Eliza Hall Institute of Medical Research, Royal Melbourne Hospital, Victoria 3050, Australia.

ment of these conditions. While health obviously improved during the period of economic development, it is clear that appropriate health intervention in the application of present knowledge could accelerate improvement in the poor countries of today as it probably would have done in Europe 300 years ago. The same argument holds for parasitic disease containment except that control measures with the efficiency of measles, polio, or pertussis vaccines have not yet been developed. Many previously successful control programs have broken down for technical or operational reasons, and in many areas from which malaria was eradicated the parasites had only at best a tenuous hold on the environment.

The cost of disease has many components, including the effect on quality of life, the cost of medical services, and the economic cost attributable to decreased productivity resulting from the morbidity and mortality of disease. Morrow, Smith, and Nimo[4] ranked the importance of disease problems in Ghana by measurement of the number of healthy days of life lost through illness, disability, and death as a consequence of disease. Using this method, premature death was rated as the most important component of the equation and malaria was ranked as the greatest disease problem. Such estimates are only as reliable as the assumptions on which they are based (and frequently the data to support those assumptions are totally lacking) and make no attempt to quantify the economic cost of disease or potential benefit of control. However, if duration and quality of life are primary determinants of the allocation of scarce resources, such an index may help in determining priorities. Similarly, government philosophy may influence the decision to allocate resources to a particular group in society for humanitarian or political reasons despite apparent economic disadvantage.

Parasitic disease most certainly demands attention for humanitarian reasons, but in this chapter we shall argue the case for increased funding for biomedical research by highlighting the fiscal benefits which could follow from the small cost of research leading to the control of disease. It will not be an emotional plea for more research but a sober attempt to review a far-from-perfect body of information on the cost of disease, to describe some of the contributions already made by basic research toward an understanding of these complex diseases, and to suggest avenues for further research. We shall not discuss the role of disease control in implementing a "Basic Needs" policy in an endemic country nor shall we compare the benefits of parasitic disease control with potential benefit of expenditure in other areas of the provision and delivery of medical care or in other areas of government expenditure such as agriculture or education. As examples of potential benefit we shall consider malaria in some detail, as well as helminth infection, schistosomiasis, filariasis, and African trypanosomiasis.

In this analysis we shall attempt to place the role of further research on parasitic diseases into realistic perspective. We hope to show that, while by no means representing the only source of progress, the new tools for prevention, diagnosis, and treatment provided through research should make a massive contribution to the eventual solution of these problems. In our view, then, research represents a cost-effective approach.

## MALARIA

There are four major species of malaria parasites which affect man. All are blood-dwelling parasites (hemoprotozoa) requiring a stage of sexual development in a mosquito to maintain the life cycle and incidentally to allow transmission within human populations. *Plasmodium falciparum* causes life-threatening infection associated with cerebral malaria, blackwater fever, and algid malaria, and is responsible for the high mortality in children in the endemic areas or nonimmune travelers acquiring the disease. Chronic or repeated infection is associated with anemia, growth retardation, and renal disease. Infection is exacerbated during pregnancy resulting in maternal morbidity and increased incidence of prematurity and low-birth-weight babies. *Plasmodium vivax* causes chronic debility with recurrent febrile illness and relapses which may occur months after the bite of an infective mosquito. *Plasmodium ovale* and *Plasmodium malariae* cause milder illness but the latter has been associated with late onset of the nephrotic syndrome thought to be a result of deposition of immune complexes in the kidney.

There are many varied estimates of the potential benefit of malaria control. Components in the different equations include effect on supply of labor through reduction in mortality, reduction in days lost through ill health, and reduction in days of decreased working efficiency caused by ill health. It is difficult enough to establish the number of deaths caused by disease but even harder to assess the number of deaths to which the disease contributed or the duration and degree of incapacity likely to be produced by the disease. Rarely is a component introduced to assess the economic cost of morbidity in children (parental care and lost educational potential) or the unemployed population, or the long-term effect on fertility rates. In fact studies by Newman in Sri Lanka (cited by Prescott[5]) suggested that the effect of malaria control in increasing birth rate was almost as great as the effect on reduction in death rate.

Pampana[6] estimates the economic loss caused by malaria in the following areas:

1. Deaths caused by malaria which result in loss of potential labor supply for the remainder of the working life.
2. Loss of earning power of individuals incapacitated by malaria. In a relapsing disease like malaria this is an important component for which a general figure is hard to establish. Components include the number of people affected and the duration and degree of disability, which obviously depend on the type of parasite and the degree of host immunity.
3. Earning power lost by the individual owing to weakness and anemia. Studies have shown that children receiving chemoprophylaxis gain more weight than untreated children, and control of malaria reduces absenteeism from school. The degree of reduction in working capacity is very difficult to assess.
4. The expenses of treatment and medical care.
5. Loss of agricultural crops due to malaria-induced shortage of labor, particularly when the peak of disease transmission coincides with the peak need for agricultural labor.
6. Death and severe illness when malaria is associated with other diseases.

7. Depreciation of land that cannot be intensively cultivated or cultivated at all on account of malaria.

Prescott[5] has reviewed many of the estimates of national output benefits of malaria eradication. Table I, reproduced from Prescott,[7] summarizes some of these conclusions and points out that their accuracy is totally dependent on the assumptions on which the calculations are based. For example, estimates for the number of days lost because of malaria vary widely in different surveys (7–44), and the degree of disability attributable to infection varies from figures such as 20% reduction in working efficiency for 6 weeks in each case[8] to 10% reduction in working efficiency throughout the year in each adult case.[9] In general there was a lack of empirical evidence to support the claims, but such figures were used as estimates for the increase in potential supply of labor to assess the potential increase in aggregate income, usually by multiplication by average national income or legislated national minimum wage.[8] Prescott notes that most studies assumed that disease control would realize these gains in aggregate income but neglected the seasonal distribution of cases, the age of the population infected, and fluctuations in the availability and seasonal needs for labor. He concluded that "No cost–benefit analysis of malaria control has yet provided a plausible estimate of the net aggregate income benefit which would result."[5]

The effect of malaria on the supply of labor and agricultural output was measured in a rural area of Paraguay by Conly.[10] In this very thorough study, three groups of farms with "low," "moderate," and "much" malaria were compared during and after a serious malaria epidemic year. Data were obtained on many indices including total harvest and number of days worked on each crop. There was a diminution in the amount of work done during the malarious year and decrease in efficiency. For certain cash crops there was an increase in labor expended but a decrease in efficiency, presumably because unskilled family members and hired laborers contributed to the total number of days worked. There was an alteration in distribution of time worked on different crops and there was lowered productivity in six of the eight crops measured and decreased efficiency of production in five of six crops studied.

**Table I.** The Output Benefits of Malaria Control[a]

| Year | Country (currency) | Scale | Years | Discount rate (%) | Deaths[b] | Disability[b] | Debility[b] | Total[b] |
|---|---|---|---|---|---|---|---|---|
| 1935 | India (pounds) | Eradication | 1 | — | 34.0 | 10.0 | 22.0 | 66.0 |
| 1959 | Philippines (US $) | Eradication | 1 | — | 15.0 | 14.0 | — | 29.0 |
| 1963 | Greece (US $) | Eradication | 1 | — | 10.5 | 12.0 | — | 22.5 |
| 1966 | Pakistan (rupees) | Eradication | 1 | — | — | 32.5 | 32.5 | 65.0 |
| 1967 | Philippines (pesos) | Eradication | 10 | — | 329.2 | 4.6 | 5.9 | 339.7 |
| 1968 | Paraguay (US $) | Eradication | 8 | 8 | — | 18.9 | — | 18.9 |
| 1969 | Iraq (dinars) | Eradication | 10 | — | 20.4 | 3.4 | — | 23.8 |
| 1971 | Thailand (baht) | Control | 15 | — | 490.1 | | — | 490.1 |
| 1971 | Thailand (baht) | Control | 15 | — | 392.2 | | — | 392.2 |
| 1975 | Sudan (Gezira) (pounds) | Control | 8 | 11 | — | 63.1 | — | 63.1 |

[a]*Source:* Prescott, N. M., 1980 (Reference 7).
[b]Millions of people.

Conly's study throws light on the potential contribution of disease control to economic development. Total crop production was reduced both for local and export markets *despite relative underemployment.* It is worthy of note that peak malaria transmission often occurs at times when the labor force is most required. Diets deteriorated to the staple manioc with decreased production of other garden vegetables. Investment by farmers in expansion and clearing of jungle for future cropping had to be reduced during the malarious year. Obviously this would lead to a decrease in potential future earnings as reduced arable land restricts the opportunities for crop rotation and adaptability for new planting of staples or cash crops. Conly comments that the ever-present threat of unstable disease prevents risk-taking for investment since the first priority should be food for home consumption. Investment in new cleared land and development of cash crops for adaptation and integration to the wider economic community will take low priority while epidemic disease prevails.

Other studies have shown that the decline in mortality following malaria eradication exceeds the fall in deaths specifically related to this disease. Giglioli[11] produced evidence from Guyana showing that infant mortality declined as did the number of deaths from malaria and chronic nephritis in children aged 1–14. There was a rapid increase in the number of live births. Vector control has the added benefit of controlling other disease spread by similar vectors, e.g., dengue fever and filariasis.

Vector control of malaria may free new areas of land for cultivation. As evidence for improvement in land value following malaria eradication, Pampana[6] cites examples from areas of Afghanistan where land price rose at least 1000-fold over a 15-year period and in a province of Cambodia where prices rose 50-fold from 1955 to 1960 with elimination of the vector. Unfortunately no data are given for price changes in comparable malaria-free regions of those countries. More relevant, perhaps, is the result produced in the Terai area of North India where malaria control programs have made available 15,000–25,000 $km^2$ of very fertile land which was uninhabitable in the 1950s.[12]

Malaria can totally disrupt the output of a major development scheme.[13] The Sennar Dam, built on the Blue Nile in 1925, transformed half a million hectares of the Gezira region of the Sudan from a sparsely populated, semi-arid savannah into a densely populated, prosperous agricultural region. In 1977 the scheme represented 12% of the total area cultivated in the Sudan but produced 75% of the country's main product (cotton) and up to 60% of the country's foreign exchange.[14] Epidemic malaria struck the area in 1950 affecting half the work force and causing hundreds of deaths. Crops were not harvested and economic loss was estimated to be $10 million. Insecticide resistance and breakdown of other vector control methods militated against disease control, let alone eradication, and by 1975 malaria was epidemic in all of 16 villages surveyed.

The difficulties of estimating the economic cost of malaria have been outlined. Some of the benefits have been mentioned and attention is drawn to the paucity of information on which to make judgments about the cost. We would reiterate the report of the 15th Expert Committee on Malaria which recommended "that WHO continue studies to develop a methodology for evaluating the social and economic effects of malaria and its eradication of control."[14] We endorse the recommendations for further studies contained in Prescott's excellent review of this subject.[5]

## THE ROLE OF MEDICAL RESEARCH IN MALARIA CONTROL

Whatever the cost of malaria, whether it be £80 million for India in 1935 or 490 million baht for Thailand in 1971, the international community has seen fit to support efforts to eradicate the disease. Bruce-Chwatt[15] estimates the cumulative expenditure on global malaria eradication from 1957 to 1977 to be not less than $2.65 billion. Given that the disease is not under control (and if anything is on the increase), there is an obligation on the scientific community to find new approaches to this problem.

The challenge to biomedical science is to improve on the results so far. In his presidential address to the American Society of Tropical Medicine in 1975 entitled "Malaria Control in the Twentieth Century," Jeffery[16] commented, "Perhaps the first step necessary in the consideration of future malaria programs, regardless of whether we call them eradication programs or control programs, is to admit in a forthright way that we have thus far failed to either achieve eradication or even to achieve a technology which is adequate for this purpose." The situation has not improved since that time. In May 1975 the World Health Assembly noted the deteriorating situation and Jeffery[16] commented that the most significant recommendation was toward "development of malaria control methodologies better adapted to the epidemiological and economic conditions in each country." One priority was the study of better methods for appropriate application of existing methods and a second was longer term research which could lead to development of new methods of control. Jeffery noted that "even an annual expenditure of 100 million dollars in 1975 would be equivalent to only about one dollar for each case, and about 100 dollars for each infant or child death attributable to malaria on the African continent alone."

The directions for research are clear. New methods of vector control and chemotherapeutic agents are necessary as insect and parasite resistance have become problems. Pant and Gratz from the Division of Vector Biology and Control of the World Health Organization have presented evidence that widespread use of insecticides in agriculture is implicated in the selection for resistant strains of Anopheline mosquitoes in India, Turkey, and Central America.[12] An interdisciplinary approach to selection of insecticides is required. Whatever method is selected, one must remember the lesson of past efforts at eradication where "one sees human behavior patterns which have militated against the effectiveness of methods used."[16] Human behavior is equally as important as any other component in the triad of host, parasite, and environment. In analyzing the feasibility of a malaria eradication program Borrelly[17] concluded that a program, to be effective, "not only requires coherence of the technical measures but also that the socioeconomic situation as a whole, as it exists and in its development dynamics, should not stand in their way. One may need to reconcile the highest target with the imagination required to find a way to reach it." What more of a challenge could be provided to behavioral psychologists or sociologists? Let us support the behavioral research which can provide imaginative ways of ensuring that the discovery of an effective tool for malaria will be

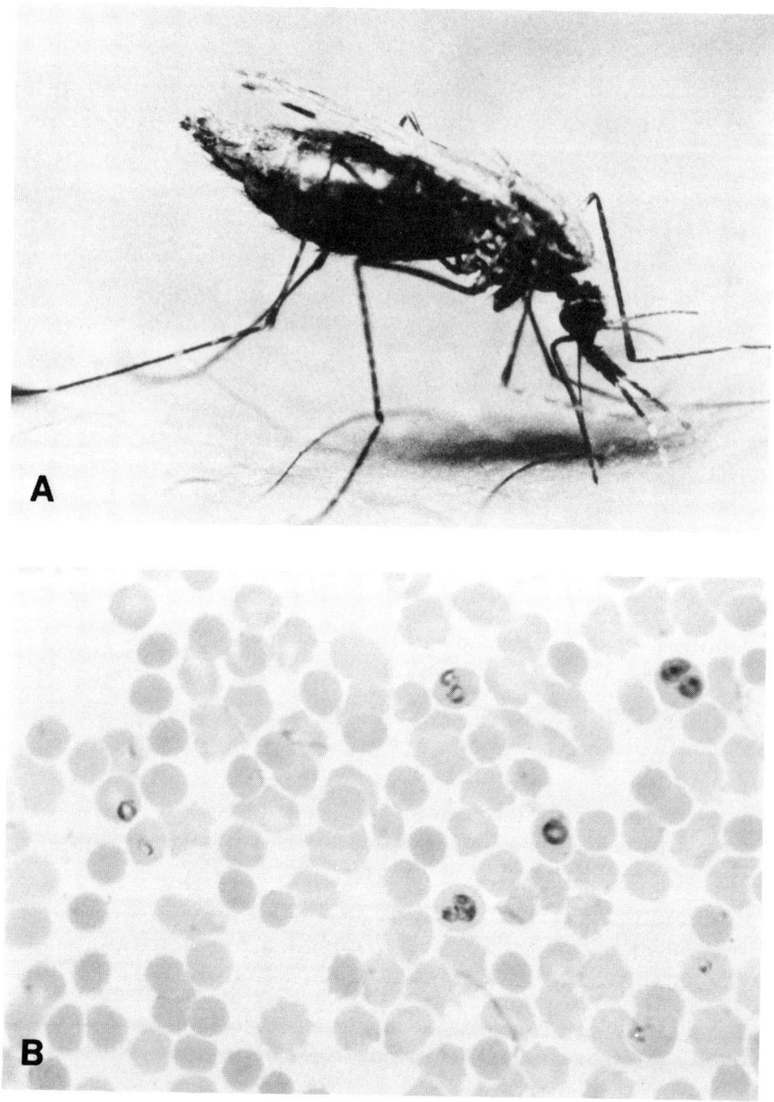

**Figure 1.** Research in malaria. (A) Mosquitoes, the vectors of malaria, are the subject of much research into methods for blocking the cycle of transmission. Photograph courtesy of the World Health Organization. (B). *In vitro* cultivation of *Plasmodium falciparum* was a major breakthrough. Parasites at various stages of maturation are seen in this illustration. (C) Two-dimensional gel electrophoresis and autoradiography provide a degree of resolution which could not be obtained with older techniques. Malaria antigens important in inducing a protective response can be identified by immunoprecipitation with immune serum.

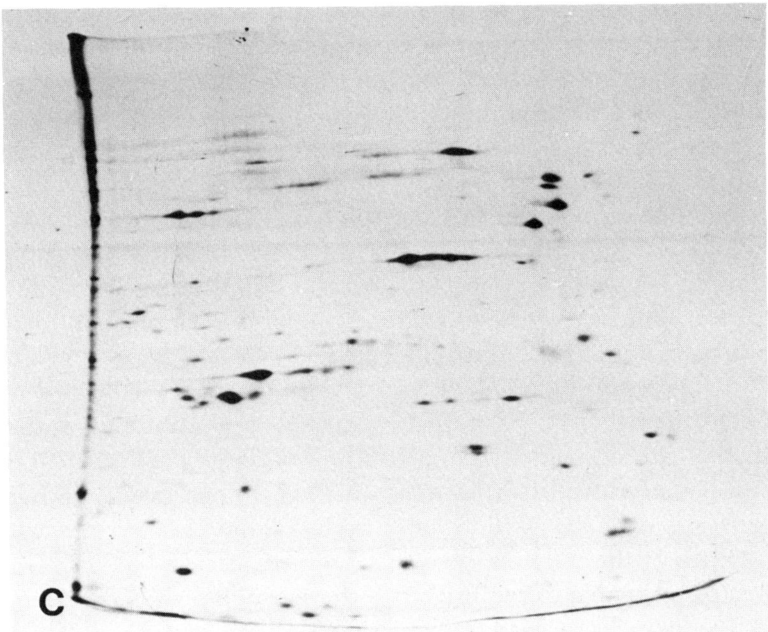


**Figure 1** *(Continued)*

more rapidly implemented than the one and one-half centuries it took to eradicate smallpox after Jenner's discoveries.

Laboratory scientists would provide a magnificent weapon by developing a vaccine against malaria, and recent results give ground for guarded optimism about the prospects. Adults living in endemic areas gradually develop immunity as shown by the ability of their serum to treat children with infection[18] and a vaccine would be designed to induce a protective immune response similar to that which develops after years of exposure to the parasite. A major hurdle was overcome when Trager and Jensen discovered a method for maintaining *Plasmodium falciparum* in long-term cultivation in the laboratory.[19] For the first time parasites could be studied in the absence of the host's immune system and large amounts of material could be obtained (Figure 1). Monkeys have been successfully immunized against infection with the human pathogen using parasites from *in vitro* culture.[20]

A major challenge to the biochemist is to identify the particular parasite molecules responsible for the induction of protection so that large quantities of a defined antigen vaccine could be prepared using techniques already available in molecular biology. A hypothetical sequence in the development of a parasite vaccine is outlined in Figure 2 and progress is well underway in this direction toward a malaria vaccine. One of the latest advances in biomedical science (hybridoma technology)[21] has already been applied to demonstrate that antibody of unique specificity (monoclonal antibody) directed against one protein on the surface of the mosquito stage of the parasite (sporozoite) can protect a mouse against challenge.[22]

Freeman and co-workers have shown that monoclonal antibody directed against antigens of the blood stages can protect mice infected with the virulent rodent parasite *Plasmodium berghei*[23] and Perrin's group has produced hybridomas which block the growth of human parasites (*Plasmodium falciparum*) *in vitro*.[24] Immunization against sexual stages of the parasite does not prevent disease but renders the parasites incapable of replication in the mosquito, thus providing an effective block to disease transmission.[25]

There is a precedent for immunization against hemoprotozoan infection. Tick fever (babesiosis) causes anemia, wasting, and death in nonimmune calves. A strain of the causative organism (*Babesia bovis*) rendered nonvirulent by passage through splenectomized calves induces infection but no disease in cattle and produces complete protection against the virulent strain of this parasite.[26] This vaccine must save many thousands of dollars in the cattle industry.

While the "angry young men" of biomedical science push toward an understanding of the biochemistry of the parasite and emphasize the hopes for a vaccine, we must not be complacent about continuing the search for total understanding of all aspects of this disease. We must take heed of the history of attempts at control of malaria and the wisdom which arose from a succession of "exaggerated expectations followed sooner or later by disappointment and abandonment of work." The only prospect of real progress lies in renewed activity in the continuous study of the disease in all its aspects.[27] In looking at the 30 years of inadequate progress in malaria control following the introduction of residual insecticides, Weller[28]

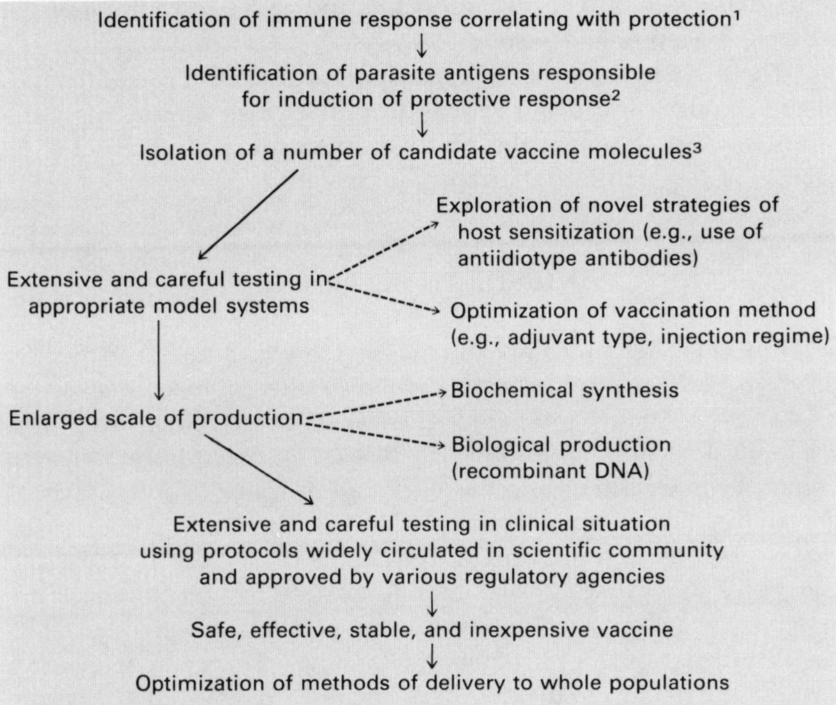


1. Preferably in natural infection but also in unnatural hosts and following immunization
2. Serum and hybridoma-derived antibodies are powerful reagents for this purpose in immunoprecipitation and binding analyses.
3. See Anders, R. F., Howard, R. J., and Mitchell, G. F., 1981, Parasite antigens and methods of analysis, *in: Immunology of Parasitic Infection and Disease* (S. Cohen and K. S. Warren, eds.), 2nd edition, Blackwell, Oxford.

**Figure 2.** Hypothetical sequence in the development of a parasite vaccine.

summarized the errors of omission including: (1) overselling the potential of eradication by residual insecticide; (2) failure to appreciate that there was inadequate understanding of the epidemiology of malaria; (3) "concentrating on a technique we lost perspective as biologists and failed to anticipate the continuum of new problems"; and (4) "captives of a technical approach we segregated ourselves from integrated programs aimed at the resolution of health problems. We must be in phase with other health programs and with concurrent levels of social and economic development."

The challenge to medical research is as strong as it was at the turn of the century, and using lessons from the past, we can apply the recently acquired knowledge and the optimism of the present to all aspects of research into the control of malaria.

## INTESTINAL HELMINTHS

Intestinal helminths are the most widespread human parasites. *Ascaris lumbricoides* (roundworm) probably infects one-quarter of the world's population[29] and has its greatest effect on children with high worm burdens in the intestine. The disease is spread from human to human by ingestion of infective eggs and its transmission is favored in communities with inadequate facilities for human waste disposal. Hookworm, though infecting fewer people (of the order of 700 million), probably causes greater morbidity. Infective larvae which develop from eggs excreted in human feces penetrate human skin, then pass via the lymphatic circulation to the lungs. From here the worms pass via the trachea to the gut to take up their final position, sucking blood from the intestinal wall to which they attach. It is estimated that hookworm causes daily blood loss throughout the world of up to 7 million liters[30]! Barefooted individuals in communities with poor sanitation (and particularly communities in which human waste is used in cultivation) have high rates of infection and disease leading to severe iron deficiency anemia.

There are few studies of the effect of ascaris on productivity of adult labor. Latham, Latham, and Basta[31] made a study in Kenya of the effects of ascaris (roundworm) infection on growth, nutritional status, and health of children under 6 years of age. They concluded that heavy infestations could lead to nonutilization of 20% of ingested calories while light infestations caused lesser reductions in absorption of nutrients. On average there is a loss of 3% of ingested calories. Treatment of infection resulted in a spurt of growth and the signs of protein calorie malnutrition were significantly reduced. Willett, Kilama, and Kihamia[32] showed in a randomized trial of treatment with antihelminthic drugs over 1 year that the rate of weight gain was 8% greater for children receiving drugs than for placebo-treated controls, and in those known to be infected with ascaris at the start of the study the rate of weight gain was 21% greater in children treated with drugs than those receiving placebo.

The treatment programs undoubtedly improved the health of children, although without social change and health education it is likely that reinfection would have occurred. Latham, Latham, and Basta[31] estimated the costs of ascaris infection in Kenya, including medication, hospitalization, and loss of income, to be *$5 million*. The major cost in this equation was the food loss of 2.8% of ingested calories.

The economic loss associated with hookworm is also attributable to decreased productivity of workers rendered anemic by this parasite. *Infection* which is mild causes no problem in comparison with *disease* brought on by large numbers of parasites causing an anemia which is likely to be much more severe when dietary iron is limited. Layrisse and Roche[33] showed that above a certain threshold of infection, hemoglobin fell as the degree of infestation increased. Basta, Soekirman, Karyadi, and Scrimshaw[34] found that 88% of adult male workers on a rubber plantation in West Java, Indonesia, had hookworm infection and 45% were anemic. Physical endurance and work performance correlated with hemoglobin. During the first study period 8% of the plantation workers suffered from infection and there was a correlation between hemoglobin and morbidity. Treatment with oral iron resulted in a significant improvement in work performance, morbidity, and endurance test capability. There was a difference in productivity of approximately 20% between anemic and nonanemic workers, and after the iron treatment the anemic group productivity reached that of the nonanemic workers.

Benefits of control of hookworm in this study were estimated from the yearly increase of latex which could be obtained by workers tapping rubber trees. The cost of iron treatment per hectare worked was 17¢ per year compared with market value of excess latex of $44. The benefit–cost ratio is calculated at 260 : 1 if it is assumed that the goods could be sold at market price and there was no limit to the supply of capital or land to extend the plantation. No estimate was made of potential future benefits of treating children and women or savings from averted disability of other infectious disease, and no attempt was made to compare the cost–benefit ratio of iron supplementation with parasite eradication. It is noted that workers had severe infection resulting in anemia and this study must be considered as a control program for relatively severe disease.

Physical endurance of sugarcane workers was found to be reduced in the presence of moderate to severe anemia[35] and correlated with decreased activity assessed by the amount of sugarcane cut per day. Figures returned to normal when daily oral iron was given to anemic workers.[36] Baker[37] and Baker and de Maeyer[38] also showed a similar reversible relationship between anemia and productivity in tea pickers.

The increased productivity associated with alleviation of hookworm anemia provides economic grounds for investing in medical research to find better ways of preventing and treating the disease. More information is required on the effect of the disease on the nutritional status of children and pregnant mothers and the relationship between maternal anemia and neonatal morbidity. Little is known of the interaction between intestinal helminthiasis and other parasitic infections such as malaria.

The immune response to hookworm infection is poorly understood but demands further study if a vaccine is to be developed. Natural infection is apparently an inadequate stimulus for induction of a protective response sufficient to cause expulsion of parasites from the gut, hence parasites persist for a long time.

Miller[39] has shown that irradiated larvae are a more potent stimulus to resistance than natural infection with hookworm in the dog, and an obvious avenue for research is the development of an appropriate laboratory small-animal model to follow up these observations. The ways in which the parasite usually evades the host response may well provide clues on the normal mechanisms of mucosal immunity in the gut as well as improving our understanding of this important disease.

## SCHISTOSOMIASIS (BILHARZIASIS)

Schistosomiasis is a chronic disease caused by small egg-laying worms living in the blood vessels of the gastrointestinal or urinary system. Eggs passed in the urine or feces hatch in fresh water and infect snails which support further development and spread of the disease. Hundreds of free-swimming larvae (cercariae) are released from the snail, and the life cycle is completed when the cercariae invade the skin, mature in the lung, and eventually reach the preferred site in blood vessels.

Schistosomiasis of the urinary system causes blood loss and bladder fibrosis, and is associated with bladder cancer. The major complication of intestinal schistosomiasis is bloody diarrhea, liver fibrosis, and altered blood flow in the liver (portal hypertension) with eventual liver decompensation and further bleeding from the bowel. Early in the course of the disease treatment may be successful, but later the changes become irreversible.

Many attempts have been made to estimate the morbidity and economic loss associated with schistosomiasis (see Table II), a disease which affects at least 200 million people.[40] In most cases the estimates are large but quite unreliable with no evidence presented concerning prevalence and severity of disease in different age groups. In reviewing the evidence Prescott[41] notes "the unrealistic assumptions made in estimating the increment in, and marginal product of, efficiency labor units which might result from control."

Several studies were performed to examine the effect of schistosomiasis on morbidity and work performance without attempting to quantitate the expected economic gain from disease control. Foster,[42] studying workers on a Tanzanian sugar estate, found increased hospital visits and inpatient admissions among infected individuals with high water exposure, who also lost more time from work than uninfected workers. Fenwick and Figenschou[43] estimated productivity by bonus earned and found that uninfected workers had significantly greater output than parasitized individuals. Treated patients improved relative to those untreated. In contrast with studies showing no effect of schistosomiasis on productivity, Prescott[41] cites

**Table II.** The Output Benefits of Schistosomiasis Control[c]

| Year | Country (currency) | Scale | Years | Discount rate (%) | Deaths[b] | Disability[b] | Debility[b] | Total[b] |
|---|---|---|---|---|---|---|---|---|
| 1949 | Egypt (pounds) | Eradication | 1 | — | — | — | 80.0 | 80.0 |
| 1951 | Egypt (pounds) | Eradication | 1 | — | — | — | 20.0 | 20.0 |
| 1958 | Egypt (pounds) | Eradication | 1 | — | — | — | 65.0 | 65.0 |
| 1958 | Egypt (pounds) | Eradication | 1 | — | — | — | 100.0 | 100.0 |
| 1967 | Egypt (US $) | Eradication | 1 | — | — | — | 500.0 | 500.0 |
| 1971 | Egypt (pounds) | Eradication | 1 | — | — | — | 213.8 | 213.8 |
| 1972 | Egypt (pounds) | Eradication | 1 | — | — | — | 45.3 | 45.3 |
| 1958 | Iraq (dinars) | Eradication | 1 | — | — | — | 6.5 | 6.5 |
| 1958 | Iraq (dinars) | Eradication | 1 | — | — | — | 10.0 | 10.0 |
| 1963 | Philippines (pesos) | Eradication | 1 | — | — | 0.4 | 1.9 | 2.3 |
| 1972 | Africa (US $) | Eradication | 1 | — | — | 173.0 | 272.8 | 445.8 |
| 1972 | Mauritius (US $) | Eradication | 1 | — | — | 0.3 | 0.5 | 0.8 |
| 1972 | S.W. Asia (US $) | Eradication | 1 | — | — | 6.7 | 9.8 | 16.5 |
| 1972 | S.E. Asia (US $) | Eradication | 1 | — | — | 64.3 | 53.8 | 118.1 |
| 1972 | Americas (US $) | Eradication | 1 | — | — | 26.4 | 34.0 | 60.4 |
| 1974 | Zanzibar (shillings) | Eradication | 60 | 10 | 21.3 | — | — | 21.3 |
| 1974 | Zanzibar (shillings) | Eradication | 60 | 15 | 10.9 | — | — | 10.9 |
| 1974 | Zanzibar (shillings) | Eradication | 60 | 20 | 6.6 | — | — | 6.6 |
| 1974 | Zanzibar (shillings) | Eradication | 60 | 25 | 4.4 | — | — | 4.4 |

[a] *Source:* Prescott, N. M., 1980 (Reference 7).
[b] Millions of people.

work from one study by Weisbrod suggesting that schistosomiasis was significantly related to lower total daily earnings among male plantation workers in St. Lucia. There was no significant difference in total weekly earnings, suggesting compensatory mechanisms such as increase in days worked or change to less demanding work. There was no correlation seen for female workers. The direct effect of severe disease on performance was investigated in the Sudan by El Karim[44] who demonstrated impaired physical working capacity (maximum aerobic power output) in physically active 18- to 45-year-old canal cleaners with heavy schistosomiasis infection. There was an impairment of 16%–18% of maximal physical working capacity in canal cleaners compared to lightly infected and uninfected villagers.

All studies mentioned have shortcomings, but suggest that at least heavy infection can affect work performance. While severe disease can reduce productivity, the aggregate loss is impossible to estimate. The threat to productivity is likely to be greatest in sick heavily infected individuals who are the least likely to have regular employment in a "captive" population suitable for study. This "silent majority" of individuals most affected by disease has been given little attention, and further research is needed on the wasted potential productivity of the self-employed and those unemployed because of ill-health. If the poor countries made welfare payments to the physically disabled, one suspects that schistosomiasis would be responsible for a large proportion of that budget in many countries.

Thus the economic loss from schistosomiasis is very difficult to assess although it is probably very high, especially among those heavily infected with the parasite. While ignorant of the economic cost, we have some idea of the financial aspects of control programs such as that in the Gezira province of Sudan. The prevalence of schistosomiasis in some areas was 50%–70% with 33,000 cases diagnosed in the province in 1977. In 1978 half of the general ward patients of the provincial hospital were bilharzia (schistosomiasis) cases. A major project in a comprehensive approach to prevention and control of water-associated diseases in the irrigated areas of the Gezira province[14] using environmental management, biological control, improved water supplies, case detection, and chemotherapy, including use of pesticides, with community participation, health education, and continuing epidemiological evaluation has had a cost estimate of $116.5 million or $5–$8 per head per year, of which approximately half is devoted to malaria, schistosomiasis, and diarrheal disease-control operations.

Another expensive control program has been in operation to help the lakeside communities which became infected after creation of the man-made Volta Lake in Ghana. In this area a research project on the epidemiology and methodology of schistosomiasis control in man-made lakes[45] had a budget of $3,827,272 between 1971 and 1978. The background information reported about this project states that although it was anticipated that schistosomiasis would probably increase as a consequence of the creation of the lake, no one expected the high prevalences observed and the rapidity with which the lakeside communities were affected. At the end of the project the annual per-capita cost of the program was $3.3—a cost which most governments simply cannot afford and would not have predicted at the time the dam was built.

Schistosomiasis can be partially controlled by present methods of snail control and chemotherapy. The methods used in locally successful programs in China may not be applicable in other parts of the world.[46] Rosenfield[47] has emphasized the importance of further study of the human element in the cycle of transmission where disease could be stopped if man never made contact with the freshwater environment of the carrier snail.

## SOME RESEARCH APPROACHES TO SCHISTOSOMIASIS CONTROL

Schistosomiasis has already cost the world too much and should be studied further in the field and in the laboratory. Worldwide funding for schistosomiasis research is an appalling 3¢ per case compared with more than $30 per case spent in the United States alone for research on muscular dystrophy.[40] Obvious goals are: more efficient diagnostic tools, better therapeutic agents for killing parasites and snails, better understanding of the natural history of the infection so that individuals at risk of developing severe disease may be detected early, and an efficient host protective vaccine.

A prototype immunodiagnostic reagent for schistosomiasis japonica in the Philippines has been developed by Mitchell and colleagues,[48] giving hope for a robust, reliable, and reproducible method of diagnosis to replace the cumbersome circumoval precipitin test now in use. The test assay is now being converted from a radioimmunoassay to an enzyme-linked immunosorbent antibody assay which would be suitable for mass screening in the field. The test depends on laboratory-derived monoclonal antibody (antibody with unique specificity) directed against a particular worm antigen. (The techniques used for making monoclonal antibodies are a direct but totally unpredicted side benefit of basic research carried out by Köhler and Milstein[21] to unravel the mysteries of antibody structure and its formation by white blood cells.) Monoclonal antibodies have a role in immunodiagnosis and also provide a means of extracting large amounts of one antigen from a complex mixture so that the properties of a single defined antigen may be tested.

The importance of laboratory models in the study of disease processes is well demonstrated in the case of *Schistosoma mansoni* in the mouse. Much information has been obtained about stage-depen-

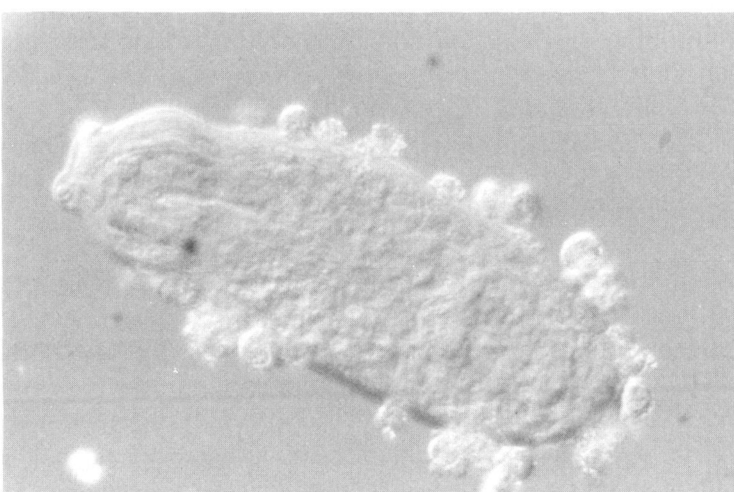

**Figure 3.** Immunopathology of schistosomiasis. Eosinophils attached to schistosomulum in the presence of antibody—the first stage in a reaction leading to parasite death. For discussion, see Reference 70 (photograph courtesy of Dr. M. Vadas).

dence of the immune response by comparing immunity to the freshly invaded schistosomula with immunity to later lung stages of the parasite. Antibody-dependent cytotoxicity of eosinophils for schistosomula has been demonstrated in the model system[49] (see Figure 3), and the mouse model enables detailed study of the immunopathology of the granulomas produced by infection.

The ultimate weapon against schistosomiasis may be a vaccine suitable for use in man. Immunization of cattle with irradiated parasites of the bovine pathogen *Schistosoma bovis* has already been field-tested in Africa and demonstrates protection induced by this procedure.[50] Identification of the antigens which produce this protection is the next step toward a defined-antigen vaccine against the disease.

## FILARIASIS

Bancroftian and brugian filariasis are the diseases caused by the parasites *Wuchereria bancrofti* and *Brugia malayi*. Larvae injected by the infective mosquitoes develop into adult worms which reside in the lymphatic glands. Microfilariae liberated by the adults circulate in the peripheral blood where they may be taken up by feeding mosquitoes. Acute infection is associated with recurrent fevers and respiratory disease while chronic infection leads to symptoms associated with longstanding lymphatic obstruction (elephantiasis and hydrocele) (Figure 4A). Basu[51] estimated that filariasis in India caused economic loss of $133 million per year associated mainly with acute disease (filarial fevers). Chronic disability, while mainly a problem in older people in Africa, may affect the working population and costs associated with surgery and convalescence must be added to the total losses.

Onchocerciasis (river blindness), transmitted by blackflies, affects over 20 million people and has devastating effects in certain communities with a large proportion of the adult population infected and up to 10% blind. In this disease the adult worms residing in subcutaneous nodules liberate large numbers of microfilariae which migrate through the skin and frequently into the eye causing an inflammatory reaction and subsequent visual impairment (Figure 4B).

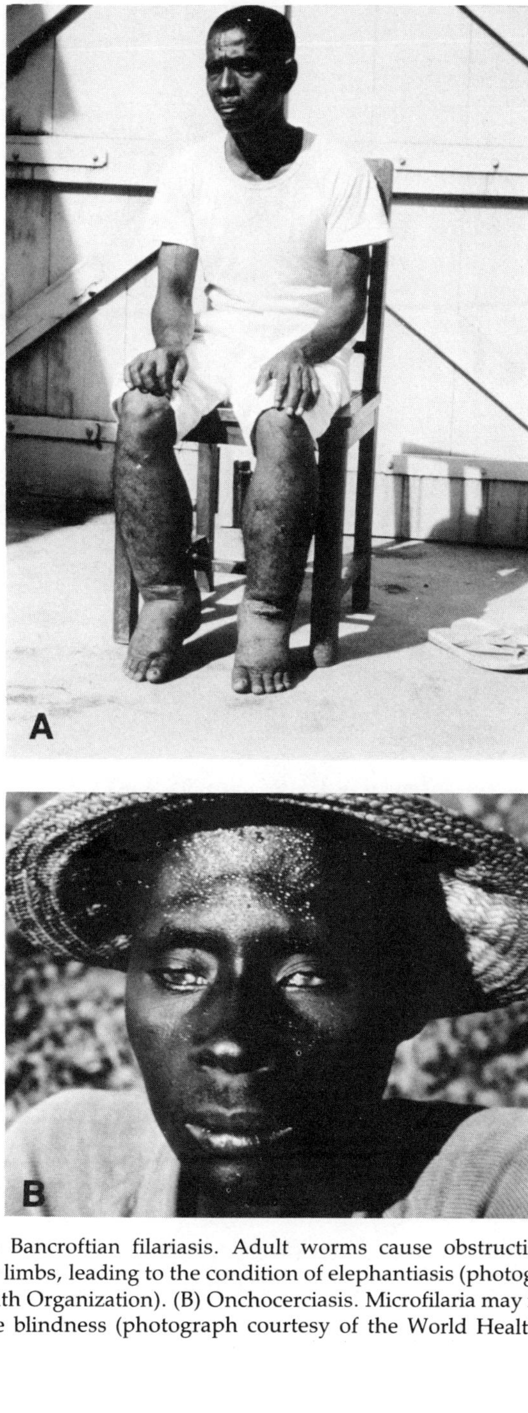


**Figure 4.** (A) Bancroftian filariasis. Adult worms cause obstruction to lymphatic drainage of the limbs, leading to the condition of elephantiasis (photograph courtesy of the World Health Organization). (B) Onchocerciasis. Microfilaria may invade the eye to cause complete blindness (photograph courtesy of the World Health Organization).

*(continued)*

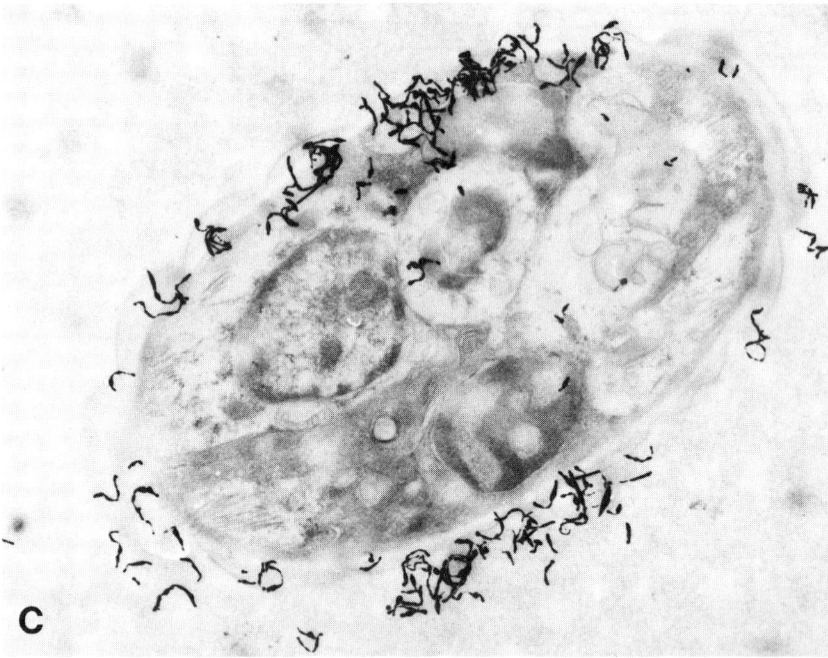

**Figure 4** *(Continued)*

(C) Filariasis research. Tangenital section of microfilaria of *Onchocerca gibsoni* showing peripheral distribution of $^{125}$I-labeled protein in an unstained electron micrograph (× 14,700, reproduced at 80%). This technique allows labeling of surface proteins for immunochemical analysis. See Reference 71.

The economic loss associated with complete blindness is difficult to assess, but for the onchocerciasis control programs the World Bank estimates showed output benefits over 25 years of $85 million or $45 million depending on the discount rate used in assessing benefits[52] (Table III). Prescott[7] has questioned the validity of some of the estimates used, but whatever the benefits happen to be, it is apparent that total eradication would yield benefit for generations to come. There is no suitable macrofilaricidal drug suitable for use in a mass program of chemotherapy.

Apart from decreased labor productivity, a major benefit of onchocerciasis control is the increased availability of formerly uninhabitable land. Rolland[53] describes the disappearance of a village fol-

**Table III.** The Output Benefits of Onchocerciasis Control[a]

| Year | Country (currency)[b] | Scale | Years | Discount rate (%) | Deaths[c] | Disability[c] | Debility[c] | Total[c] |
|---|---|---|---|---|---|---|---|---|
| 1970 | Ghana (cedis) | Eradication | 15 | 10 | — | — | 4.9 | 4.9 |
| 1970 | Upper Volta, Ivory Coast Mali (CFAF) | Eradication | 1 | — | 1 | 170.0 | — | 170.0 |
| 1978 | OCP Area (US $) | Eradication | 25 | 10 | — | 50.1 | 35.0 | 85.1 |
| 1978 | OCP Area (US $) | Eradication | 25 | 15 | — | 26.3 | 18.6 | 44.9 |
| 1978 | OCP Area (US $) | Eradication | 50 | 3 | — | 1137.6 | | 1137.6 |
| 1978 | OCP Area (US $) | Eradication | 50 | 7 | — | 368.3 | | 368.3 |
| 1978 | OCP Area (US $) | Eradication | 50 | 8 | — | 285.3 | | 285.3 |
| 1978 | OCP Area (US $) | Eradication | 50 | 9 | — | 233.4 | | 233.4 |
| 1978 | OCP Area (US $) | Eradication | 50 | 10 | — | 198.8 | | 198.8 |
| 1978 | OCP Area (US $) | Eradication | 50 | 15 | — | 100.5 | | 100.5 |

[a] *Source:* Prescott, N. M., 1980 (Reference 7).
[b] OCP = Onchocercias is Control Program.
[c] In millions of people.

lowing the development of hyperendemic onchocerciasis, while Bradley[54] concludes that disease was an important factor causing population migration away from an endemic area. The Onchocerciasis Control Program in West Africa (OCP) will make available 276,000 $km^2$ of land of which about 154,000 $km^2$ would be useful for agriculture.[55]

Thus control of onchocerciasis would lead to increased productivity and significant increases in land available for cultivation. The OCP has been financed to the extent of $60 million for the period 1974–1979 and anticipated further funding for 1980–1985 is $133 million[55] in the belief that major benefit will ensue to the 10 million inhabitants protected. Since vector control and chemotherapy are the only weapons available at present, there will always be a requirement for constant, costly surveillance to prevent resurgence of disease. Vector resistance and pollution of water sources may eventually retard the program. If biomedical science could contribute by providing better chemotherapeutic agents, better understanding of the parasite vectors and their habitat, or a vaccine against the infecting parasite, the economic cost of this disease could be reduced beyond measure. As Prof. George Nelson has pointed out,[56] "It is unlikely that there will be any significant advance in the control of filariasis and onchocerciasis as a result of strengthening normal health services." Later he states, "If we are to accelerate programs for the control of these diseases, we need more research to find new tools." The specific areas he suggests for examination are epidemiology, characterization of parasites, diagnosis, vectors, chemotherapy, and immunology (Figure 4C). At present there are no satisfactory laboratory models for the human diseases and this deficit is a major impediment to research in all forms of human filarial disease.

## AFRICAN TRYPANOSOMIASIS

African trypanosomiasis causes economic loss by its effect on people, cattle, and reduction in availability of land. An individual bitten by an infected tsetse fly develops a systemic illness, then major manifestations related to central nervous system involvement by a meningoencephalitis progressing to the terminal symptoms of sleeping sickness. The true incidence of disease is at least 10,000 new cases per year[57]; to the costs of expensive treatment campaigns must be added the decreased productivity of infected individuals and years of labor lost through premature death of adult workers. The disease in cattle is a factor retarding economic growth by its effect on reduction in supply of milk and meat to human populations, and Nash[58] has concluded that it "is of far greater economic importance than sleeping sickness; further it is of much greater medical importance, if the broad view is taken that a deficiency in animal protein among the masses is far more important than a specific disease among the few." Like onchocerciasis, trypanosomiasis and its vector, the tsetse fly, are important factors in causing migration of populations with their animals away from infected areas,[59] and disease control in heavily infested areas of Africa would lead to a significant increase in available land. The economic losses caused by this group of diseases have not been estimated, but there is a clear need for their documentation.

## FASCINATING RESULTS IN TRYPANOSOMIASIS RESEARCH

Infection with African trypanosomes is associated with waves of parasitemia in the peripheral blood and each wave is serologically distinct from the previous ones, even if a single parasite is used to initiate infection in a laboratory animal. It has now been shown that the major component of the trypanosome coat is a variant specific glycoprotein,[60] and each wave of parasitemia is associated with a different antigenic type. The ability of the parasite to switch from one antigenic type to another (antigenic variation) provides a very effective means of evading the host's protective immune response. Syringe-passage of trypanosomes into a nonimmunized host is associated with reversion to a basic antigenic type,[61] and a similar phenomenon occurs with transmission through the tsetse fly. The molecular biology of this process has been the subject of much recent study[62–64] to determine how the genomic rearrangement occurs to allow expression of different variant specific genes. The challenge for the immunologist is to find a molecule or method of immunization which induces an immune response superseding antigenic variation which allows the host to respond to any surface antigenic variant produced by the parasite.

## SMALLPOX ERADICATION: AN ECONOMIC SUCCESS

The success of the smallpox eradication campaign program in terminating for all time the ravages of this disease is now well known. The final thrust of this international effort depended on local and regional cooperation to apply the principles of epidemic surveillance to eradicate the last remaining foci. With no known nonhuman host and a very effective vaccine, the disease was finally conquered nearly 200 years after Jenner's observations gave rise to an effective weapon against the infection. Lest we become too optimistic about vaccines, let us remember that the situation is far less favorable for parasitic disease where no vaccine is presently available for any of the major pathogens. All countries of the world are threatened as international migration and travel transmit diseases between countries. In 1979, for example, 894 cases of malaria were reported in the United States, a 22% increase over 1978, and five deaths were reported in the same year.[65]

The saving in health expenditure following the success of the smallpox eradication program is less well documented and publicized. It is estimated that a total amount of $313 million was spent in eradicating smallpox, or an average of $24 million per year for the period 1967–1980. It has been conservatively estimated that routine vaccination programs, quarantine services, control of international travelers, and medical care of complications of vaccination cost the world about $1 billion per year.[66] The developed countries of the world which supported this program have all benefited in their own right by dismantling of quarantine and vaccination facilities. In 1968 smallpox-related activities cost the United States government $150 million while United States support for the eradication program was 6% of this amount.[67]

## HOW MUCH DO WE SPEND ON RESEARCH AND WHO PAYS?

In 1979 it was estimated that there was approximately \$40–\$50 million available worldwide for basic and applied research on all tropical diseases[40]—a pitiful sum considering that this money could not buy even two F-14 jet fighter planes[67]! Scientific progress toward better understanding of the disease processes so they may be better controlled demands greater allocation of research funds from all countries of the world. In 1978 it was estimated that 97% of world funding on biomedical research was spent on health problems that affect the rich countries while only 3% was spent on diseases that are the scourge of the poor.[68] Organizations like the National Institutes of Health (NIH) have no authority to work on health problems of people in other countries. Research on tropical diseases accounts for only 0.06% of the total NIH budget[67] (1979 estimate).

In the absence of research initiative from the rich countries of the world, the World Health Organization, World Bank, and United Nations Development Project have shown great leadership in setting up the Special Programme for Research and Training in Tropical Diseases which is described in another chapter of this volume. Another major contributor to this worldwide problem has been the Rockefeller Foundation, and one cannot underestimate the contribution of its International Health Division to the understanding of all aspects of malaria, hookworm, and yellow fever epidemiology and control in the three decades from 1920 to 1950.

Foundations continue to provide major support for this neglected area of medical research. Some, such as the Edna McConnell Clark Foundation, concentrate on all aspects of one disease, in their case, schistosomiasis. Of special interest is the program of the Rockefeller Foundation under whose auspices a network of laboratories has been established to support research into the "great neglected diseases of mankind."[68] The Foundation aims to create links between biomedical research workers and laboratories wherever they may be in the world so that the latest information from the field and the latest developments of biomedical science can be applied together to the most pressing disease problems of the world today. The Foundation has also provided a number of "Career Development Awards" to enable promising young scientists to continue a career in an area of biomedical science which is normally very poorly supported.

## FUNDAMENTAL INTEREST OF PARASITISM

For economic reasons we can justify further research aimed at conquering parasitic disease in areas where the diseases are endemic and for selfish and economic reasons we could justify preparing ourselves to deal appropriately with diseases introduced to nonendemic areas by tourists and migrants, but it is the fundamental interest of the questions being asked which could attract many scientists into this field. How do parasites evade the immune response? How do living parasites reside in macrophages? What is the controlling factor in genomic reorganization in trypanosomiasis? How is drug resistance induced? Undoubtedly the study of parasitic disease will increase our general understanding of biology and may even give clues to mechanisms of prevention of cancers and other conditions which are extremely rare in populations affected by endemic parasitic disease.[69]

## CONCLUSION

Parasitic diseases are a major cause of morbidity and mortality, and present methods have not been successful in controlling or eradicating them to improve the quality of life of afflicted populations. Economic costs of these diseases are extremely large but difficult to assess, and further research is necessary to quantify the extent to which they retard economic development. Evidence has been presented that potential productivity of populations would be increased to a great extent by disease control, and in some instances there would be significant increase in supply of land and improved nutrition through control of animal disease.

New approaches are required to supplement the present inadequate tools at our disposal to deal with parasitic disease and the relatively small cost of increased funding for medical research is likely to be balanced by large long-term benefits. The excitement generated by the study of parasitic disease and the promising progress of the past few years ensures that further commitment of basic and applied biomedical research to the major parasitic diseases of man, combined with the application of techniques already developed, will lead to better methods of diagnosis and treatment which will greatly reduce the social and economic impact of these diseases.

The target of health for all by the year 2000 demands more than idealism for its achievement. The attempt to attain this laudable goal imposes an obligation to strive for more than the application of present knowledge coupled with continuing epidemiological evaluation. Biomedical scientists must be given the opportunity to take up the challenge to devise futuristic weapons with which to counteract the persistent onslaught of parasitic disease.

## ACKNOWLEDGMENTS

The authors wish to thank Graham Mitchell, Andrew Abbot, and Sonya Belan for helping with illustrations and reviewing the manuscript. The authors are supported by grants from the National Health and Medical Research Council of Australia, the Rockefeller Foundation Great Neglected Diseases of Mankind Network, and the UNDP/World Bank/World Health Organization Special Programme for Research and Training in Tropical Disease, and Grant Number AI-03958 from the National Institute of Allergy and Infectious Diseases, U.S. Public Health Service.

## REFERENCES

1. Bloom, B. R., 1979, Games parasites play: how parasites evade immune surveillance, *Nature* **279:**21–26.
2. *Malaria Processed Report for the Special Programme for Research and Training in Tropical Diseases,* 1976, World Health Organization, Geneva.
3. *Health,* Sector policy paper, 1980, World Bank, Washington, D.C., p. 14.
4. Morrow, R. H., Smith, P. G., and Nimo, K. P., 1980, A quantitative method of assessing the health impact of different diseases in less developed countries, Report of UNDP/World Bank/WHO Special Program for Research and Training in Tropical Diseases, TDR/SER/SWG(2)/80 WP3a.

5. Prescott, N. M., 1980, The economics of malaria, filariasis and human trypanosomiasis, Report of UNDP/World Bank/WHO/Special Program for Research and Training in Tropical Diseases, TDR/SER(SC-1)/80.4.
6. Pampana, E., 1969, The cost of malaria and its eradication, *in: A Textbook of Malaria Eradication,* Oxford University Press, New York, Chapter 17.
7. Prescott, N. M., 1980, On the benefits of tropical disease control, *in: Health Policy in Developing Countries,* Royal Society of Medicine International Congress and Symposium Series No. 24 (C. Wood and Y. Rue, eds.), Royal Society of Medicine and Academic Press, London, Grune & Stratton, New York (U.S. edition), pp. 43–45.
8. San Pedro, C., 1967, *Philipp. J. Public Health* **12:**5, quoted in Prescott, N. M., 1980, On the benefits of tropical disease control, *in: Health Policy in Developing Countries,* Royal Society of Medicine International Congress and Symposium Series No. 24 (C. Wood and Y. Rue, eds.), Royal Society of Medicine and Academic Press, London, Grune & Stratton, New York (U.S. edition).
9. Sinton, J. A., 1935, *Rec. Malar. Surv. India* **5:**223 and 413, and **6:**96, quoted in Prescott, N. M., 1980, On the benefits of tropical disease control, *in: Health Policy in Developing Countries,* Royal Society of Medicine International Congress and Symposium Series No. 24 (C. Wood and Y. Rue, eds.), Royal Society of Medicine and Academic Press, London, Grune & Stratton, New York (U.S. edition), pp. 43–45.
10. Conly, G. N., 1975, *The Impact of Malaria on Economic Development: A Case Study,* Pan American Health Organization, Pan American Sanitary Bureau, Regional Office of the World Health Organization, Scientific Publication No. 297.
11. Giglioli, G., 1972, Changes in the pattern of mortality following the eradication of hyperendemic malaria from a highly susceptible community, *Bull. WHO* **46:**181–202.
12. Pant, C. P., and Gratz, N. G., 1979, Malaria and agricultural development, *Outlook Agric.* **10**(3)**:**111–115.
13. *Tropical Diseases,* 1976, World Health Organization, Geneva.
14. Plan of operations for a comprehensive approach to the prevention and control of water-associated diseases in the irrigated schemes, Gezira Province, Sudan, Report of UNDP/World Bank/WHO Special Program for Research and Training in Tropical Diseases, EM/SUD/VBC/001, 1979.
15. Bruce-Chwatt, L. J., 1978, Chapter 5, *in: Tropical Medicine: From Romance to Reality* (C. Wood, ed.), Academic Press, London.
16. Jeffery, G. M., 1976, Malaria control in the twentieth century, *Am. J. Trop. Med. Hyg.* **25:**361–371.
17. Borrelly, R., 1976, A method of analyzing the socioeconomic aspects of the feasibility of malaria control programmes, Progress Report of UNDP/World Bank/WHO Special Program for Research and Training in Tropical Diseases, WHO IR/ARI/007.
18. Cohen, S., McGregor, I. A., and Carrington, S. C., 1961, Gammaglobulin and acquired immunity to malaria, *Nature* **192:**733–737.
19. Trager, W., and Jensen, J. B., 1976, Human malaria parasites in continuous culture, *Science* **193:**673–675.
20. Reese, R. T., Trager, W., Jensen, J. B., Miller, D. A., and Tantravahi, R., 1978, Immunization against malaria with antigen from *Plasmodium falciparum* cultivated *in vitro, Proc. Nat. Acad. Sci. USA* **75:**5665–8.
21. Köhler, G., and Milstein, C., 1975, Continuous cultures of fused cells secreting antibody of pre-defined specificity, *Nature* **256:**495–497.

22. Yoshida, N., Nussenzweig, R. S., Potocnjak, P., Nussenzweig, V., and Aikawa, M., 1980, Hybridoma produces protective antibodies directed against the sporozoite stage of malaria parasite, *Science* **207**:71–73.
23. Freeman, R. R., Trejdosiewicz, A. H. J., and Cross, G. A. M., 1980, Protective monoclonal antibodies recognizing stage-specific merozoite antigens of a rodent malaria parasite, *Nature* **284**:366–367.
24. Perrin, L. H., Ramirez, E., Lambert, P. H., and Miescher, P. A., 1981, Inhibition of *P. falciparum* growth in human erythrocytes by monoclonal antibodies, *Nature* **289**:301–303.
25. Carter, R., and Chen, D. H., 1976, Malaria transmission blocked by immunization with gametes of the malaria parasite, *Nature* **263**:57–60.
26. Callow, L. L., and Mellors, L. T., 1966, A new vaccine for *Babesia argentina* infection prepared in splenectomized calves, *Aust. Vet. J.* **42**:464–465.
27. *Second General Report of the Malaria Commission of the League of Nations Health Organization*, 1927, quoted in Jeffrey, G. M., 1976, Malaria control in the twentieth century, *Am. J. Trop. Med. Hyg.* **25**:361–371.
28. Weller, T. H., 1974, World health in a changing world, *Am. J. Trop. Med. Hyg.* **77**(Supplement):54–61.
29. Wilcocks, C., and Manson-Bahr, P. E. C. (eds.), 1972, *Manson's Tropical Diseases*, Bailliere Tindall, London, p. 247.
30. Brown, H. W., 1975, *Basic Clinical Parasitology*, 4th edition, Appleton-Century-Crofts, New York.
31. Latham, L., Latham, M., and Basta, S. S., 1977, The nutritional and economic implications of ascaris infection in Kenya, World Bank Staff Working Paper No. 271, World Bank, Washington, D.C.
32. Willett, W. C., Kilama, W. L., and Kihamia, C. M., 1979, *Ascaris* and growth rates: A randomized trial of treatment, *Am. J. Public Health* **69**:987–991.
33. Layrisse, M., and Roche, M., 1964, The relationship between anemia and hookworm infection, *Am. J. Hyg.* **79**:279–301.
34. Basta, S. S., Soekirman, M. S., Karyadi, D., and Scrimshaw, N. S., 1979, Iron deficiency anemia and the productivity of adult males in Indonesia, *Am. J. Clin. Nutr.* **32**:916–925.
35. Davies, C. T. M., 1973, Relationship of maximum aerobic power output to productivity and absenteeism of East African sugar cane workers, *Br. J. Ind. Med.* **30**:146–154.
36. Davies, C. T. M., and van Haanen, J. P., 1973, Effect of treatment on physiological responses to exercise in East African industrial workers with iron deficiency anaemia, *Br. J. Ind. Med.* **30**:335–340.
37. Baker, S. J., 1978, Nutritional anaemia—a major controllable public health problem, *Bull. WHO* **50**(5):659–675.
38. Baker, S. J., and de Maeyer, E. M., 1979, Nutritional anaemia: Its understanding and control with special reference to the work of the World Health Organization, *Am. J. Clin. Nutr.* **32**(2):368–417.
39. Miller, T. A., 1966, Comparison of the immunogenic efficiencies of normal and x-irradiated *Ancylostoma caninum* larvae in dogs, *J. Parasitol.* **52**:512.
40. Hoffman, D. B., 1979, *Schistosomiasis Research: The Strategic Plan*, Edna McConnell Clark Foundation, New York.
41. Prescott, N. M., 1979, Schistosomiasis and development, *World Dev.* **7**(1):1–14.

42. Foster, R., 1967, Schistosomiasis on an irrigated estate in East Africa: III. Effects of asymptomatic infection on health and industrial efficiency, *Am. J. Trop. Med. Hyg.* **70:**185–195.
43. Fenwick, A., and Figenschou, B. H., 1972, The effect of *Schistosoma mansoni* on the productivity of cane cutters on a sugar estate in Tanzania, *Bull. WHO* **47:**507–572.
44. El Karim, M. A. A., 1980, An assessment of physical performance in Gezira population infected with schistosomiasis, World Health Organization TDR/SER/SWG(2)/80 WP2.
45. Research on the epidemiology and methodology of schistosomiasis control in man-made lakes: Project findings and recommendations, Report prepared for the Governments of Ghana and Egypt by the World Health Organization acting as Executing Agency for the United Nations Development Programme, 1979, PDP/79.2, United Nations Development Programme and the World Health Organization, Geneva.
46. Bruijuing, C. F., 1980, Control of schistosomiasis in the Peoples' Republic of China, *in: Health Policy in Developing Countries,* Royal Society of Medicine International Congress and Symposium Series No. 24 (C. Wood and Y. Rue, eds.), Royal Society of Medicine and Academic Press, London, Grune & Stratton, New York (U.S. edition), pp. 43–45.
47. Rosenfield, P. L., 1980, Schistosomiasis transmission and control: The human content, Report of UNDP/World Bank/WHO Special Program for Research and Training in Tropical Diseases, TDR/SER/(SC-1)/80.5.
48. Mitchell, G. F., Cruise, K. M., Garcia, E. G., and Anders, R. F., 1981, A hybridoma-derived antibody with immunodiagnostic potential for schistosomiasis japonica, *Proc. Nat. Acad. Sci. USA* **78:**3165–3169.
49. Butterworth, A. E., David, J. R., Franks, D., Mahmoud, A. A. F., David, P. H., Sturrock, R. F., and Houba, V., 1977, Antibody-dependent cell-mediated damage to $^{51}$Cr-labeled schistosomula of *Schistosoma mansoni:* Damage by purified eosinophils, *J. Exp. Med.* **145:**136–150.
50. Majid, A. A., Bushara, H. O., Saad, A. M., Hussein, M. F., Taylor, M. G., Dangie, J. D., Marshall, T. F. de C., and Nelson, G. S., 1980, Observations on cattle schistosomiasis in the Sudan, a study in comparative medicine: Part 3: Field testing of an irradiated *S. bovis* vaccine, *Am. J. Trop. Med. Hyg.* **29:**452–455.
51. Basu, P. C., 1971, Interpretation of eight years data in the control of *W. bancrofti* filariasis in urban areas of India by vector controls using conventional antilarval measures, *J. Parasitol.* **57:**57–58.
52. World Bank, 1978, Onchocerciasis control programme: Economic review mission, unpublished document, World Bank, Washington, D.C., cited in Prescott, N. M., 1980, On the benefits of tropical disease control, *in: Health Policy in Developing Countries,* Royal Society of Medicine International Congress and Symposium Series No. 24 (C. Wood and Y. Rue, eds.), Royal Society of Medicine and Academic Press, London, Grune & Stratton, New York (U.S. edition), pp. 43–45.
53. Rolland, A., 1972, Onchocerciasis in the village of Saint Pierre: An unhappy experience of repopulation in an uncontrolled endemic area, *Trans. R. Soc. Trop. Med. Hyg.* **66**(6)**:**913–915.
54. Bradley, A. K., 1976, Effects of onchocerciasis on settlement in the Middle Hawal Valley, Nigeria, *Trans. R. Soc. Trop. Med. Hyg.* **70:**225–229.
55. Bazin, M., 1980, The economic aspects of the onchocerciasis control programme in the Volta Basin, *in: Health Policy in Developing Countries,* Royal Society of Medicine

International Congress and Symposium Series No. 24 (C. Wood and Y. Rue, eds.), Royal Society of Medicine and Academic Press, London, Grune and Stratton, New York (U.S. edition), pp. 31–32.

56. Nelson, G. S., 1980, Research as an aid to filariasis and onchocerciasis control, *in: Health Policy in Developing Countries,* Royal Society of Medicine International Congress and Symposium Series No. 24 (C. Wood and Y. Rue, eds.), Royal Society of Medicine and Academic Press, London, Grune & Stratton, New York (U.S. edition), pp. 167–172.
57. De Raadt, P., 1976, African sleeping sickness today, *Trans. R. Soc. Trop. Med. Hyg.* **70**(2)**:**114–116.
58. Nash, T. A. M., 1960, A review of the African trypanosomiasis problem, *Trop. Dis. Bull.* **57:**973–1003.
59. McConnell, E., and Baker, J. R., 1970, A further look at sleeping sickness in Ethiopia, *Trans. R. Soc. Trop. Med. Hyg.* **64:**162–163.
60. Cross, G. A. M., 1975, Identification, purification and properties of clone-specific glycoprotein antigens constituting the surface coat of *Trypanosoma brucei, Parasitology* **71:**393–417.
61. Capbern, A., Giroud, C., Baltz, T., and Mattens, P., 1977, *Trypanosoma equiperdum:* étude des variations antigeniques au cours de la trypanosomose experimentale du lapin, *Exp. Parasitol.* **42:**6–13.
62. Hoejimakers, J. H. J., Borst, P., Van den Burg, J., Weissman, C., and Cross, G. A. M., 1980, The isolation of plasmids containing DNA complementary to messenger RNA for variant surface glycoproteins of *Trypanosoma brucei, Gene* **8:**391–417.
63. Hoeijmakers, J. H. J., Frasch, A. C., Bernards, A., Borst, P., and Cross, G. A. M., 1980, Novel expression-linked copies of the genes for variant surface antigens in trypanosomes, *Nature* **284:**78–80.
64. Williams R. O., Young, J. R., and Majiwa, P. A. O., 1979, Genomic rearrangements correlated with antigenic variation in *Trypanosoma brucei, Nature* **282:**847–849.
65. Annual Summary, 1979, Reported morbidity and mortality in the United States, *Morbidity Mortality Weekly Rep.,* 1980, **28**(54), U.S. Department of Health and Human Services, Atlanta.
66. The global eradication of smallpox, 1980, WHO, Geneva, p. 64.
67. Schultz, M. G., 1980, The forgotten problems of forgotten people, *in: Health Policy in Developing Countries,* Royal Society of Medicine International Congress and Symposium Series No. 24 (C. Wood and Y. Rue, eds.), Royal Society of Medicine and Academic Press, London, Grune & Stratton, New York (U.S. edition), pp. 57–62.
68. Warren, K. S., and Romney, H., 1978, Tropical medicine revisited, *The Rockefeller Foundation Illustrated* **4**(2).
69. Oomen, A. P., 1980, African lessons for Western medicine, *in: Health Policy in Developing Countries,* Royal Society of Medicine International Congress and Symposium Series No. 24 (C. Wood and Y. Rue, eds.), Royal Society of Medicine and Academic Press, London, Grune & Stratton, New York (U.S. edition), pp. 31–32.
70. Vadas, M. A., Butterworth, A. E., Sherry, B., Dessein, A., Hogan, M., Bout, D., and David, J. R., 1980, Interactions between human eosinophils and schistosomula of *Schistosoma mansoni, J. Immunol.* **124:**1441–1448.
71. Forsyth, K. P., Copeman, D. B., Abbot, A. P., Anders, R. F., and Mitchell, G. F., 1981, Identification of radioiodinated surface proteins and antigens of *Onchocerca gibsoni* microfilariae, *Acta Trop.* **38:**329–342.

# 8

# The World Health Organization
## Its Influence on Worldwide Research Policies

H. C. GOODMAN and T. A. LAMBO

The World Health Organization (WHO) is the specialized agency of the United Nations dealing with health. WHO implements the policies established by its member countries at the annual World Health Assembly in Geneva, Switzerland. WHO's research program originated therefore because its member countries resolved that their health organization should promote research to solve those problems "which are better resolved by world cooperative endeavour than by local groups" and also to solve "major problems when local effort is inadequate. . . ."[1] Interestingly—and perhaps unexpectedly—WHO's research program, from its modest beginning in 1958, now has a pronounced feedback effect on the biomedical research policies of its member nations. A detailed account of the genesis of WHO's research program has been published by Corning.[2] This chapter will describe the development of WHO's research program by illustrating the impact on national research policies of WHO's recent emphasis on regional research development and especially on the concept of "special programs," using as an example the Special Programme for Research and Training in Tropical Diseases.

The constitution of WHO has always required the Organization to "promote and conduct research in the field of health." WHO's

H. C. GOODMAN • Tropical Medicine Center, Johns Hopkins University, Baltimore, Maryland 21205. T. A. LAMBO • World Health Organization, Geneva, Switzerland.

formal research program, however, was initiated in its 10th year as a result of a proposal made by the delegation of the United States to the 11th World Health Assembly in 1958. A special contribution of $300,000 by the U.S. financed a study requested by the Assembly which led to the formulation of the role of WHO could play in the promotion and coordination of worldwide medical research and on intensification of WHO's promotion of medical research.

The conclusion by leaders in medical research meeting at WHO was that its role in international research should concern itself primarily with:

1. The solution of those problems which are better resolved by world cooperative endeavor than by local (i.e., national) groups.
2. The solution of major problems when local effort is inadequate either because the research potential of the area is limited or because of the nature of the problem.[1]

Because WHO's financial resources for research promotion have always been limited, its funds have served as "seed money" to catalyze research in national institutions of its member nations on subjects arising from Assembly resolutions by member nations.

Table I compares the research funds available to WHO in the

**Table I.** Comparison of Biennial WHO Research Budgets of 1970–1971 and 1980–1981[a]

| | Total regular budget[b] | Regular budget: research | "Extrabudgetary" research[c] |
|---|---|---|---|
| 1970 + 1971 | 141,125 | 8,200 (5.8%) | 9,452 (4,820 IARC) |
| 1980 + 1981 | 427,290 | 10,300 (2.4%) | 122,300 (14,798 IARC) (54,607 TDR) (37,500 HRP) |

[a]$ in 1000s.
[b]Regular budget: annual assessments from member countries.
[c]"Extrabudgetary": contributions by member countries (or donations by outside agencies) in excess of annual assessments, usually earmarked for specific projects. IARC, International Agency for Research on Cancer; TDR, Special Program for Research and Training in Tropical Diseases; HRP, Special Program for Research, Development, and Research Training in Human Reproduction.

1970–1971 biennium with the 1980–1981 biennium. The regular budget research funds—from annual assessments of member countries—rose only slightly in the decade and fell in percent (from 5.8% to 2.3%) of the regular budget. The budgets of the International Agency for Cancer Research (IARC) in Lyons and of the two "Special Programmes"—for research and training in tropical diseases (TDR) and for research, training, and development in human reproduction (HRP)—account for $107 million of the $122 million in *extrabudgetary* (special contributions and donations) research support for 1980 and 1981.

How can a research budget for all of WHO of $17.7 million in 1970 and 1971 or of $54.6 million for tropical diseases in 1980 and 1981 fund the research to solve the world's major health problems when in one country, the U.S., over $1 billion is expended annually on research on cancer alone? The answer clearly is that the WHO funds are not intended to support the totality of the research to be undertaken, and WHO, by working through national institutions whose personnel, equipment, and overhead costs are met by its member countries, has catalyzed global research programs mobilizing national resources amounting to many times the actual funds allocated for research in its annual budget. One striking example of the success of this policy is the WHO/United Nations Development Programme/World Bank-sponsored Special Programme for Research and Training in Tropical Diseases, which was initiated to meet the problems of tropical diseases discussed in Chapter 7 and which is described later in this chapter.

## THE "SPECIAL PROGRAMME" CONCEPT

The "Special Programme" concept was developed in WHO as an attempt to resolve the paradox of the existence of a global health research expenditure of over $10 billion at the same time that health problems in developing countries remain overwhelming and, for example, chloroquine-resistant malaria is spreading in Asia and Latin America and schistosomiasis is on the increase in tropical countries. The paradox is simply explained: 95% of health research takes place in industrialized countries and is directed at the health problems of those countries.[3]

"These [special] programmes were conceived and developed by WHO in response to the urgent need of developing countries to find their own solutions to their health problems. The transfer and application of technologies produced for the industrialized countries have frequently turned out to be inappropriate, even counter-productive. The same has been true for the training of scientists from developing countries in the latest scientific techniques of the industrial world. The WHO is in a uniquely advantageous position to promote and coordinate research and training.

"The prototype for the Special Programmes—'The Expanded Programme of Research, Development and Research Training in Human Reproduction'—began in 1971. This programme adopted the concept of goal-oriented research and of systems of scientific peer review. Most importantly, it placed both the scientific planning and the execution of the programme in the hands of national scientists and institutions, through the mechanisms of task forces and contract research. In 1974 and 1975 these concepts were developed further during the planning phases of the Special Programme for Research and Training in Tropical Diseases, . . . [which] represents the most advanced stage of evolution of the Special Programme concept . . . and the two programmes have evolved together into a bold new attempt to harness scientific resources on a global scale to help the developing countries find solutions for their health problems."[4]

Tropical diseases are unlike smallpox which posed a continuing threat of disastrous outbreaks in the developed world. The industrialized countries readily joined with WHO and the countries in which smallpox was endemic to fund the WHO smallpox-eradication program. In fact smallpox was not the highest priority problem in many developing countries, but nevertheless these countries invested proportionately huge resources in money and manpower in the successful collaborative eradication effort organized through WHO. The eradication of smallpox saves the industrialized world over $1 billion which had been spent each year for prevention or expensive quarantine measures following each imported case of smallpox.

As is the case in other developed countries, the $3-billion annual budget for health research in the U.S. is, as one might expect, directed toward the priority health problems of its citizens. A generous estimate is that less than 2% of that amount supports research on tropical diseases (see Table II).[5] And yet prevention and control of the parasitic and other tropical diseases and diarrheal diseases obviously rank high on the list of *global* health priorities.

By pursuing twin goals, the special programs of WHO provide a framework for the international collaborative research required to solve health problems of the developing countries while at the same time providing the framework for international collaboration in research training to assist the developing countries to increase their research capability. The twin goals are:

1. Multidisciplinary and multinational research to provide a better understanding of health problems and to develop new approaches and new tools for the control of disease and health promotion.
2. The strengthening of biomedical research capabilities of the developing countries so they are enabled to deal more effectively with their own health problems.[3]

**Table II.** Estimated Support for Parasitology Research, FY 1979[a]

| Organization | Cost |
|---|---|
| U.S. government agencies | |
| USPHS (including NIH) | $16,365 |
| USAID | 10,200 |
| U.S. Army | 11,584 |
| U.S. Navy | 1,995 |
| U.S. Department of Agriculture | 7,028 |
| Center for Disease Control (CDC) | 1,400 |
| GORGAS | 227 |
| National Science Foundation (NSF) | 33 |
| Total government support | $48,831 |
| Private foundations | |
| Rockefeller Foundation | $ 2,515 |
| Clark Foundation | 2,176 |
| Total foundation support | 4,691 |
| Total support | $53,522 |

[a]Modified from Cook, J. E., 1981 (Reference 5); $ in 1000s.

The two goals are not separable; the disparity between the biomedical research resources of industrialized versus developing countries is so great that any effort to form a partnership for multinational collaborative research is doomed to failure unless a training component is an integral part of each research program and project. The developing countries need the biomedical research know-how of the industrialized countries, and the industrialized countries want to make their expertise available for solving global health problems. But it is no longer possible, as it was in the days of colonial empires, to solve problems by sending in teams of researchers. Partnership in collaborative research is a requirement for success.

In 1964 WHO initiated the concept of encouraging and supporting research and training centers in developing countries with the pioneer WHO Immunology Research and Training centers in developing countries. In 1970 the guiding aims and principles of the centers were reviewed in a Round Table Conference on the Training of Research Workers in the Medical Sciences,[6] and these same principles have been incorporated into the "Special Programmes" described below. Essentially an institution or academic department is identified which has the potential for developing the type of scientific research required to solve national or regional priority health problems. Scientists from other regions are brought to the center to help in training and collaborative research. By carrying out initial training of scientists in their own country or region, training abroad can be for shorter periods and in the specific techniques and disciplines needed for progress of research on problems of local importance. The ultimate goal is the development of a center of experience and knowledge for national or regional research and training which can carry out collaborative research with institutions in developed countries on an equal partnership basis.

## WHO AND THE PROMOTION AND COORDINATION OF NATIONAL HEALTH RESEARCH IN THE DEVELOPING COUNTRIES

In industrialized countries one central body or often several agencies, both governmental and nongovernmental, are usually responsible for defining health priorities and research objectives. However, in developing countries for the most part there was no effective national organization for the promotion, coordination, and/or management of national health research. This was one of the main reasons why WHO recently added five additional regional research advisory committees to the original Headquarters (now global) and Pan-American Health Organization Advisory Committees on Medical Research. Annual meetings of representative health scientists, and especially those representing existing or emerging national medical research councils (or their equivalents), in each of the six regions of WHO have confirmed the need for stronger promotion, coordination, and management mechanisms for national health research planning and implementation in the developing countries. WHO has not only provided this framework of meetings to catalyze the process, but has supplied consultants as well. In Thailand, for example, at the request of the Thai government, WHO helped sponsor five meetings of government and university biomedical and health-delivery researchers, and during a visit by a consultant, mechanisms for national health research planning were discussed, including the strengthening of the existing medical research subcommittee of the National Research Council. Recommendations were made for the Ministry of Health to bring together working groups of government health workers and university researchers to plan the laboratory and field research required for control of chloroquine-resistant malaria, diarrheal diseases, and others of Thailand's major health problems. In each region of WHO the need and the opportunity for annual meetings with other national health research planners has resulted in a reevaluation in all developing countries of the planning, coordination, and management mechanisms for health research.

## THE SPECIAL PROGRAMME FOR RESEARCH AND TRAINING IN TROPICAL DISEASES

In his annual report to the World Health Assembly in 1971, Dr. Candau, then director-general of WHO, described the frustration of national health authorities in the face of the worsening situation with regard to malaria, schistosomiasis, filariasis, and African and American trypanosomiasis, and stressed the need for more research in chemotherapy, vector control, and immunology (for both diagnosis and prevention):

"From this analysis of the situation it is clear that the control of these diseases—to which many other examples could be added—is unlikely to make much progress in the next 10 or even 20 years unless significant discoveries provide new means of attack. The best hope for the future, therefore, and the most rational long-term approach, lies in a vigorous intensification of the search for new knowledge through carefully planned and well co-ordinated research programmes aimed specifically at finding solutions to the difficulties at present confronting health authorities and public health workers. We know what needs to be done and what questions to ask, and we have the technical resources and the research skills to find the answers. Practically 25 years after the establishment of the World Health Organization—an institution set up to enable countries to exchange technical information on all matters concerning health and to pool their resources for the common good—the time has come to apply the lessons of the past and to devise new strategies for dealing with a number of fundamental public health problems."[7]

Shortly thereafter several separate events led WHO to provide the leadership required to intensify the search for the new knowledge required by health workers faced with the responsibility for controlling tropical diseases. In 1974 Dr. Kaunda, president of Zambia, during a visit by one of us (T.L.), offered the resources of the Ndola General Hospital to WHO as a regional center for clinical laboratory and field research in tropical diseases. Both authors had already discussed the pros and cons of proposing to Dr. Mahler, a new director-general, the formation of a Division of Immunology, Genetics, and Molecular Biology to promote tropical disease research. In addition one of us (H.G.) was appointed in 1973 to the Board of Trustees of the International Laboratory for Research on Animal Diseases (ILRAD). This institution was to be (and has been) constructed in Kenya to develop a vaccine against African trypanosomiasis and theileriosis (east coast fever), because control of these two animal diseases would allow the doubling of beef production in Africa. The sponsorship of ILRAD, as well as what was then seven other agricultural research institutions, was by the Consultative Group for International Agriculture Research (CGIAR) with its secretariat cosponsored by the Food and Agriculture Organization of the United Nations (FAO), the United Nations Development Programme (UNDP), and the World Bank, and with additional financial support from the agencies for technical cooperation and foreign aid of the industrialized countries. The logic was clear: WHO's director-general stated that one of WHO's major concerns was the health of the tropical world's populations who would need to produce the grains, tubers, and animals developed by application of the research findings from the agricultural institutes of the CGIAR. In 1974 the Executive Board and the World Health Assembly endorsed the director-general's proposals for intensifying research on tropical diseases and for promoting research in developing countries and the strengthening of research and training centers in these countries. The global Advisory Committee on Medical Research approved plans for a Special Programme for Research and Training in Tropical Diseases (TDR) in its annual meeting in June 1974. The director-general appointed a tropical disease research group to staff the program whose "goals are to obtain effective new vaccines, diagnostic tests, drugs, and measures for vector control through research and development and concurrently to assist the tropical countries to improve their own research."[8]

Because of the limited WHO research budget, funding was expected to be from extrabudgetary sources. In preparing proposals for potential donors, many procedures were followed which had proved useful in WHO's first large extrabudgetary research project, the Expanded Programme for Human Reproduction (HRP). HRP was initiated in 1972, and following the formation of TDR in 1974, HRP in return was renamed the Special Programme of Research, Development and Research Training in Human Reproduction. One major contribution of HRP which was incorporated into the TDR program was the key decision that multidisciplinary task forces or scientific working groups and their respective steering committees be responsible for the research and development activities. They define the research objectives, devise a strategic plan to achieve them, and review the plan and the research as the work progresses:

"The research and development operations of the [TDR] Programme focus upon six tropical diseases: malaria, schistosomiasis, filariasis (including onchocerciasis), trypanosomiasis (both African sleeping sickness, and the American Chagas' disease), leishmaniasis and leprosy. These diseases were selected by the criteria of their public health importance in the developing countries, the lack of effective tools for their control, and the likelihood that research can yield the new tools which could bring significant advances in their control. The research therefore ranges from the most sophisticated laboratory investigations to the use of simple diagnostic tests in the field, and it interlinks with related sectors such as nutrition, economics and education. This area of the Programme is also concerned with the establishment of an epidemiological and operational basis for the application of improved and new tools. The research programme is essentially implemented by collaborating institutions and their scientists, e.g., out of over 500 active projects only three are operated by WHO."[3]

Research on tropical diseases sponsored and catalyzed by WHO dates back to the beginning of WHO's research program. In the malaria-eradication program, for example, it soon became apparent that research was needed to assess resistance of mosquitos to insecticides, and in immunological methods to simplify case detection, in epidemiology, chemotherapy, and vector biology. The limited WHO research funds often provided an indispensable initial impetus for work which would subsequently be funded from national resources. One can trace back the origins of present British Medical Research Council support for development of a malaria vaccine to the WHO grant made to Guys Hospital in 1967 for the initiation of pioneer work on malaria immunology in Dr. Sydney Cohen's laboratory. Today the TDR Scientific Working Group on Malaria uses its funds to intensify existing research in many countries to pursue the promising leads described under research in malaria in Chapter 7.

The development of the leprosy research program is an excellent example of how WHO action focused national research interest and support in developed countries on the need for research to develop control measures for a disease which affects over 10 million people in the less developed countries. No vaccines exist for any of the six major tropical diseases included in the WHO TDR program. The Leprosy and Immunology units of WHO had jointly arranged meetings of scientific groups on the immunology of leprosy in Geneva in 1970 and in New Delhi in 1972. Leprosy, the only nonparasitic disease of the six TDR diseases, suddenly became a prime candidate for vaccine development because of the discovery that *M. leprae,* the causative agent, which has never been grown in culture (*in vitro*), could produce massive infections when injected into armadillos. This 1971 discovery from Louisiana made possible the harvesting of large numbers of leprosy bacilli for research.

The background of how the first Scientific Working Group of TDR was organized to deal with the Immunology of Leprosy (IMMLEP) with financial support from the Norwegian Agency for International Development has been published by Dr. Tore Godal. Dr. Godal, chairman of the Steering Committee of the IMMLEP Scientific Working Group (SWG), described the workings of a SWG and the strategic plan for development of a vaccine as the "best estimate by the best people at any one time."[9] The aim was to assure that any scientist wishing to collaborate had access to the strategic plan and was continuously informed about the SWG activities, and to set up a peer group selection process for projects to be supported by IMMLEP not only on the basis of scientific merit but also (and most important) in *relationship to the goals of the program.* When asked whether the scientific working group approach is a good way to promote research on tropical diseases, Dr. Godal answered, "This is a general and basic question very much debated in research management. Many scientists especially are critical of this approach. A major basis for their concern is that this kind of organised research tends to monopolise ideas on how to get around a problem, preventing maximum diversity of approaches. This criticism might be more relevant in fields where the resources are vast and gaps in knowledge at a basic level extensive, than in areas where the research has been neglected, the resources available limited, and where the world is polarised into one part where the diseases are found and a second part where the research capability is more advanced. In this situation, which is that of leprosy and some other tropical diseases, I think that a scientific working group approach has advantages, such as recruitment of funds into the area, facilitation of international collaboration, distribution of scarce materials and active recruitment of research capability. Another argument for the goal orientation of a SWG, in the case of tropical diseases, is that otherwise there could be an increased danger of diverting the funds to the most prestigious research centres in the world, rather than to research most relevant to, for example, leprosy, since the projects then could only be selected on the basis of scientific merit."[9]

The annual reports of the TDR program document gratifying progress toward the goals of the strategic plan with shipments of infected armadillo tissue and purified *M. leprae* to 23 investigators in 1980–1981. A first batch of killed *M. leprae* for possible use in human vaccine studies is under preparation at the Wellcome Laboratories in England. The work of the network of laboratories in North and South America, Western Europe, the USSR, and Japan, which has led to this point is described in the annual reports and by Lucas.[10]

Through June 1981 a total of 1178 projects in tropical disease research have been funded by TDR for a total of $58.5 million. Some of the promising leads in research on malaria, schistosomiasis, filariasis, and trypanosomiasis are mentioned in Chapter 7. Countries making financial contributions, as well as tropical countries making contributions in kind, are represented on a Joint Coordinating Board, and, following reports by the Scientific and Technical Advisory Committee and other evaluation mechanisms, a budget of $27.5 million has been proposed for 1981. For WHO, the executing agency, this amount represents the largest of its research programs (see Table I for the *biennial* budget for research supported from extrabudgetary sources).

The following section will deal with the impact of the TDR research program on the research policy of the United States, as an example of one of WHO's industrialized member countries which did not consider tropical diseases as one of its priorities for health research.

## THE INFLUENCE OF THE WHO TDR PROGRAM ON RESEARCH POLICIES IN THE U.S.

It should be emphasized again that WHO research policies result from decisions of its member countries. The U.S. government gave support to the TDR concept in the World Health Assembly from the first, and the late John Knowles, then president of the Rockefeller Foundation, had raised that foundation's voice in a plea for increased research in the area of tropical diseases.[11] A representative of the Edna McConnell Clark Foundation had worked with WHO in the early days of planning the TDR program, and representatives of the United States Agency for International Development (USAID), the Rockefeller Institute, and the Edna McConnell Clark Foundation attended the first meeting of potential donors in November 1974. The following account will, however, focus on the WHO feedback effect on and reinforcement of U.S. initiatives to intensify interest in and increase U.S. resources for research on tropical diseases.

An important milestone was passed on February 18, 1976, at hearings of the Subcommittee on Health of the U.S. Senate. An American immunologist, Dr. Silverstein, participated in the initial TDR Scientific Working Group meeting on filariasis. He was coincidentally working with the Senate Subcommittee on Health during a sabbatical year as a Public Affairs Fellow of the Federation of American Societies of Experimental Biology. He brought the WHO initiative to the attention of Sen. Ted Kennedy, who scheduled public hearings on February 18, 1976, on the extent of the problem of tropical diseases and what the U.S. was then doing, and ought to be doing, to help solve them. The subcommittee invited testimony from: WHO on the Special Programme for Research and Training in Tropical Diseases; Dr. Theodore Cooper, then assistant secretary for health of the Department of Health, Education, and Welfare (DHEW, now the Department of Health and Human Services); Dr. Paul Ehrlich, director of DHEW's Office of International Health; the director of the Office of Health of USAID; Dr. Richard Krause, the director of the National Institute of Allergy and Infectious Diseases (the institute responsible for research in tropical diseases of the National Institutes of Health); as well as from representatives of the Center for Disease Control (Dr. William Foege) and the Food and Drug Administration.

Senator Kennedy's opening statement was an excellent summary of both the problems and the opportunities: "The health problems of many countries, and peoples—particularly in the developing world—are staggering. Three infectious diseases alone afflict half a billion people throughout the world. The number of people infected at any given time by malaria, filarial worms, or by the blinding diseases, exceeds the entire population of the United States. In Africa alone, malaria kills some 1 million children a year, and in parts of Africa 1 person in 10 is sightless from river blindness.

"These diseases and their consequences impose heavy burdens upon the world. Both disease and the fear it brings cause large areas of otherwise productive land to go unworked. Illness saps the strength of its victims. It impedes improvements in productivity and progress in economic development. It condemns millions of people to live partial lives as burdens on society instead of productive useful citizens.

"Here in the United States we have solved most serious problems of infectious diseases, and with significant American help, the age-old horror of smallpox will soon be wiped from the face of the earth. Now, it is time, therefore, for us to double and redouble the attack on other significant disease problems.

"Nor would this attack on disease require major economic resources. At the level of health care that is needed, it is not an expensive fight, but the value in human health for every dollar spent is remarkable.

"In my view, the United States, as the richest and most advanced nation in the world, has a humanitarian obligation to the rest of the world to help solve these serious disease problems. Until the end of the 1960s, we Americans had a good record on foreign assistance, including aid to health programs. But, recently, this aid has decreased substantially, both as a percentage of our gross national product, and even in absolute amounts. As was pointed out by the recently published report of a study mission to Asia and the Middle East undertaken by representatives of my Subcommittee on Health and Refugees, American foreign aid is now at a scandalously low level, and is now 14th out of 17 among the developed countries in terms of percentage of GNP.

"Even beyond purely humanitarian motives, there are strong reasons why we should support—and lead a worldwide battle against disease. In direct terms, no American traveler abroad, or even an American at home, can feel fully protected from the spread of these infectious diseases. Any contribution we make to improving the health of the world's people must inevitably contribute to accelerated economic development and prosperity, a reduction in national and international tensions, and an enhancement of the prospects for peace in many parts of the world.

"If we are to continue our generous tradition of sharing our advantages with others in the form of foreign aid—as indeed we should—then we must continuously reassess the form which that aid takes. We should be wise enough to recognize that aid in the area of international health, directed toward the solution of important disease problems can have highly significant and lasting effects—which may make humanitarian assistance to developing countries much more directly in our own interest than is generally appreciated.

"It is not merely a question of more dollars. We in the United States have the world's leading biomedical research establishment. It is the knowledge derived from our impressive medical research progress, and the broad scientific and professional talent that we have here in abundance we must share with the rest of the world.

"The purpose of this hearing is to document as fully as possible, not only the extent of the infectious disease problems that trouble millions of the world's people, but also what the United States is currently doing, and ought to be doing, to help solve them."[12]

Evidence that the hearings had the desired effect on the U.S. government research structure is evident from the testimony before the subcommittee. Dr. Cooper, speaking for DHEW, said: "Our efforts have been both bilateral with other countries and multilateral through formal relationships. The multilateral effort has concentrated in our participation in the World Health Organization and the Pan American Health Organization. . . .

"The numbers of people affected worldwide by these tropical diseases are immense. Some of the finest research activities in this category are conducted at the National Institutes of Health itself, within their own intramural programs. We are supporting roughly $7.7 million worth of research in filariasis, malaria, trypanosomiasis, schistosomiasis, leishmaniasis, Hansen's Disease, and the like, in this regard. We think that we have the technology and the scientific base for expanded effort. We hope the interest in this field in this country would be expanded."[12]

Dr. Krause said: "The NIH research support for malaria is $1.25

million, the bulk of that in the Institute of Allergy and Infectious Diseases. The majority of this, although by no means all, is centered on the fundamental biology of the organism in an attempt to probe two different areas. The first is concerned with the way in which the malaria parasite establishes infection and produces disease. The second area relates to the ways in which the body develops immune processes against malaria. In understanding both of these fundamental areas we hope that new avenues of drug development on the one hand, or vaccine development on the other, would be developed for the eventual control of disease.

"It seems to us that there are many instances where [environment control] will be difficult to achieve. We therefore must, at the same time, push ahead with control through either new drug development or through new vaccine development.

"NIH support in grants and contracts, not only for malaria, but also for all of the six diseases which have been highlighted by WHO, is about $6.2 million at this time in grants and contracts, which represents about 5 percent of the budget of the Institute of Allergy and Infectious Diseases. . . . There are, however, additional activities that we have not been doing, and which we will try to do. I think we should begin to look at program projects in parasitic diseases which would pull together specialists in cell biology and molecular biology—two new powerful ways of looking at disease problems. These program projects would focus on parasitic diseases."[12]

The WHO testimony by one of us (H.G., then director of the Special Programme for Research and Training in Tropical Diseases), gave the rationale for the need for the U.S., as a WHO member country, to implement the program which WHO had initiated in response to the delcared wishes of its member countries, including the U.S. The testimony emphasized the need for a three-pronged approach for progress in health: economic development, the better application of existing knowledge, and the development of new technologies. Each of these is indispensable, yet none of them is sufficient in itself; together they form an interlocking triad for health promotion:

"In recent years, the third element has been greatly neglected in comparison with the efforts put into economic development and the application of available methods. This is not meant as a denigration of past efforts, but rather as an indication that their scale has matched neither the increasing needs for improved tools nor the great potential for advances.

"WHO together with the United Nations Development Program has recently established a Special Program for Research and Training in Tropical Diseases. This is not a bureaucratic activity to be carried out behind closed doors. Scientists and health workers from industrialized and tropical countries are being sought together to begin to plan, conduct, and supervise the research through to the stage where effective remedies are produced.

"These scientists will work through a network of existing laboratories situated both in tropical and industrialized countries to carry out the research and to train more researchers. New laboratories will be created only if they are found to be necessary.

"To make a beginning, we will initially emphasize six diseases for which present remedies are inadequate and on which insufficient research is now in progress. These are malaria, schistosomiasis, filarial infections, trypanosomiasis, leprosy, and leishmaniasis. Work has already begun through an international task force to develop a leprosy vaccine.

"We cannot now predict when a program on this scale may end, but its duration is foreseen in terms of decades. It is the Organization's view that ultimately the responsibility for research on these diseases must lie with the tropical countries themselves, and that a key element in the program must therefore be the development of self-reliance in research by these countries.

"As to cost, an annual operational budget of the order of $15 million is expected in the initial stages of the program. This special program has been established in response to the declared wishes of the Organization's member countries, including the United States, as expressed in resolutions of the World Health Assembly. The Organization looks to its member countries to implement the program, both through their technical resources and by contributions to WHO's Voluntary Fund for Health Promotion.

"The technical resources of the United States for research of this type are enormous. For example, those of the National Institutes of

Health. WHO very much hopes that research in tropical diseases in the United States can be linked to the real problems of tropical countries throughout the Organization's Special Programme. A crucial effort is now being started to develop powerful technical advances for the control of tropical diseases, and it is gratifying that the Senate Health Subcommittee has considered this as a matter of high priority."[12]

Less than two weeks following the hearings Senator Kennedy sent a list of 11 questions to the representatives of each of the U.S. government agencies who testified at the hearings. Included in the questions were the following:

"The World Health Organization has initiated a special program for research and training in tropical diseases. How do you see your department interfacing with this new WHO effort?"

"What are the obstacles to real breakthroughs in the cure of tropical diseases?"

"What additional resources have you identified to further research in tropical diseases of this sort?"[12]

The hearings in February 1976 and Senator Kennedy's list of questions were important factors in stimulating the executive branch of the U.S. government to undertake reassessment of U.S. activities in international health and health research, a reassessment which has continued over the past 5 years. A detailed account of U.S. initiatives in international health, and especially of the Carter administration years, can be found in Corning's *Review of the U.S. Role in International Biomedical Research and Communications.*[2] The role of WHO in influencing these initiatives is summarized below. Peter Bourne, the White House advisor for health after the election of President Carter in 1976, attended WHO World Health Assemblies and was an important administration spokesman for an increased role of the U.S. in international health. President Carter wrote a letter to the president of the World Health Assembly endorsing WHO initiatives,[13] including the tropical diseases program, during a visit he made to Geneva in 1977. Further, in Caracas in May 1978 President Carter said: "I am announcing today my intention to launch a program to strengthen the participation of the United States in worldwide efforts to overcome disease and ill health. We will work toward the lasting control or eradication of the major infectious diseases; particularly 'tropical' diseases that continue to be the leading causes of death and disability.

"We will give special attention to the major causes of death in children—diarrhea and respiratory disease, particularly when aggravated by malnutrition.

"We plan to:

"—Work closely with nations around the world, individually and through organizations such as WHO, UNICEF, the World Bank, and the regional development banks, to improve the health of people everywhere.

"—More fully involve American universities, technological foundations, and other private organizations in making U.S. scientific and professional resources more accessible to the developing world.

"I have asked Secretary Califano to go to Geneva as head of the United States delegation to the World Health Assembly, where he will describe further the steps my Administration intends to take. . . .

"The strategy I am announcing today and which Secretary Califano will describe to the World Health Assembly next week will be developed as part of our budgetary and legislative recommendations for next year. Where possible, however, we will move immediately to carry out this program."[14]

In Geneva a week later, Joseph Califano, then secretary of Health, Education, and Welfare, addressed the World Health Assembly in these words: "It is my honor to speak today as the first Cabinet officer ever to head the United States delegation to the World Health Assembly. I come as President Carter's personal emissary, to underscore the commitment of the government and people of the United States to the World Health Organization.

"So I come today, at the President's behest, to announce several initiatives of my government in the field of international health: efforts to be launched in partnership with the World Health Organization and other nations; efforts to be undertaken in the spirit of interdependence which makes the World Health Organization a model of international cooperation.

"First, we want to commit new resources to the battle against infectious diseases. The United States is already deeply involved in combatting the major infectious diseases. We support the World

Health Organization's Special Program of Training and Research in the Tropical Diseases; we pursue research in our own institutions; we collaborate on disease control projects in individual countries.

"To support the Tropical Disease Research Program over the next five years we have recently pledged more than $20 million. We are particularly eager, through this work, to help countries develop their own capability in research and training.

"Studies of the tropical diseases in our own research institutions include not only the six diseases in the TDR Program, but several others, including cholera and the diarrheal diseases. . . .

"We recognize, however, that these existing efforts are not enough. So we are increasing our training and research in tropical diseases. We have established a post-doctoral fellowship program in cooperation with the World Health Organization, which will bring scholars from other countries to study tropical diseases in our institutions.

"To help improve the research capability of countries where tropical diseases are a problem, we are also launching new programs of international cooperation. Our National Institutes of Health will, for example, also help establish and strengthen research centers in developing countries.

"We believe that we must do more in the future if we are to have an impact on the enormous world health problems that confront us. Accordingly, President Carter has directed that over the next few months we forge plans for more effective use of our resources, private and governmental.

"—We will seek, therefore, to increase support for our universities and institutions, and for institutions in developing countries to strengthen their capabilities for research, training, and the effective delivery of health services.

"—We will seek to increase support for training, for both our own people and those from developing countries, who can work as research scientists, educators, and providers of health care.

"—We will seek to increase support for international activities of our own government agencies whose competence should be more readily available to international health activities.

"We are, as you know, a nation seeking to reassert our commitment to human rights—at home and in the world. Our participation in this organization gives us an opportunity to express that commitment not simply in words, but in deeds."[15]

Similar pledges were made for U.S. contributions both of funds and of national health research resources to the goals of other developing WHO programs: the Expanded Programme on Immunization (to deliver vaccines presently available), Nutrition, Prevention of Blindness, Primary Health Care, and the United Nations Decade for Drinking Water and Sanitation.

Meanwhile in the legislative branch Senator Kennedy appeared before the WHO Medical Society in Geneva in 1977 where he singled out two WHO programs—primary health care and research on tropical diseases—for commendation. He followed up on pledges made at the February 1976 hearings by introducing in March and June 1976 a series of successful amendments to the foreign assistance (AID) program to strengthen and expand U.S. policies and programs in the international health field. He referred to tropical diseases in his introduction to the ammendments as follows: "Why is not more being done? Why do the experts tell us that the situation will get worse by the year 2000? It is time that the people of the United States and other advanced nations rise up in outrage at these conditions, and demand that governments work together to end this global blight. In the United States, very little research is being carried out on tropical diseases, and yet this must be an important component of any foreign assistance program.

"Mr. President, I would suggest that not only can much be done about these disease problems, but that dollars spent in this area are some of the most cost effective in terms of foreign assistance. Further, improvements of the health of peoples abroad can have a direct influence on the health and the health-care costs of the American people. Again, I need only cite one example of this. The U.S. contribution to the worldwide smallpox eradication campaign cost us a total of some $25 million, but because eradication of this disease enabled us to terminate our own domestic smallpox vaccination program, we are now able to save over $125 million a year domestically as the direct result of our modest international effort in this area."[16]

Both he and Senator Jacob Javits conducted hearings and introduced broadly based international health bills which would have provided the funds by U.S. institutions to develop linkages with developing country institutions for research and training in tropical diseases. Unfortunately these bills were introduced just as Congress adopted a policy of cutbacks in government spending and they were never adopted.

Of the pledge in 1978 of $20 million over 5 years to what is now the UNDP/World Bank/WHO Special Programme for Research and Training in Tropical Diseases (TDR), $10.3 million has been contributed through fiscal year 1981 (it is of interest to note that through June 1981, 220 projects were funded to investigators in the U.S. totalling $11.8 million). In addition the National Institute of Allergy and Infectious Diseases (NIAID) has authorized the Fogarty Institute to establish postdoctoral fellowships in U.S. institutions for training TDR-sponsored applicants from institutions in developing countries in tropical diseases research. Moreover in 1979 NIAID established a program for International Collaboration in Infectious Disease Research (ICIDR) to link U.S. institutions to research units at overseas sites. Actually the ICIDR was implemented with funds originally budgeted for the International Centers for Medical Research which terminated in 1980, although the replacement focused the research support more sharply on the goals of the WHO TDR program. Finally a new NIAID program for Tropical Disease Units in U.S. institutions was established in 1980. A Tropical Disease Research Unit would be expected to develop programs of basic and applied research in a multidisciplinary attack on the world's tropical and parasitic diseases. Unfortunately funding restrictions for the budgets of the National Institutes of Health has allowed only two such units to be funded to date.

In 1979 the Carter administration prepared a proposal for a mechanism to assist less developed countries help themselves to apply science and technology (on a broader scale than health) to problems of development. As part of a reorganized governmental foreign aid structure, an Institute for Scientific and Technological Cooperation (ISTC) was proposed with a budget of $25 million in 1980 and $40 million in 1981. The authorization bill for the ISTC was passed by both Houses; however the U.S. legislative process requires a subsequent appropriation bill to authorize funding. In the end no funds were appropriated and the ISTC died. However, a more modest beginning of a program to meet the goals of sharing U.S. science and technology with developing countries has been implemented through a reorganization at AID under the Reagan administration which will create a new high-level Bureau for Science and Technology which will administer AID grants to universities in the U.S. and abroad to foster technological links between the U.S. and developing countries. Nominated for director with the rank of senior associate administrator is Nyle Brady, formerly director of the CGIAR's International Rice Institute in the Philippines.[17] In addition the National Academy of Sciences will receive a $36-million grant from AID to support science and technology in developing countries. It is estimated that about half of the 5-year grant, to be administered by the National Academy of Science's Board on Science and Technology for International Development, will be used to fund research and development projects in developing countries. While the major funding will be for agricultural and other fields of science and technology, a health committee has been appointed and a portion of the funding will be available for tropical disease research.

## CONCLUSION

The relationship between the establishment of WHO research policy by its member nations and the subsequent feedback effect on the health research policy of its member nations is well illustrated by the UNDP/World Bank/WHO Special Programme for Research and Training in Tropical Diseases. A framework for international collaboration and coordination is provided through WHO's execution of the program with its budget of $27.5 million for 1981. It is self-evident that this amount cannot finance the physical structures, the laboratory equipment, the overhead, and the training and employment of the skilled scientific manpower required to find new prevention and control measures for tropical diseases. The investment in these elements by the tropical countries themselves is massive in relationship to their national income. For effective implementation of the institutional collaboration between industrialized and tropical countries, however,

WHO also depends on the commitment of resources by the industrialized countries. In the U.S., at a recent Macy Foundation Conference on the Present Status and Future of Parasitology, U.S. support for research in parasitology in 1979 was estimated to be $53.5 million (Table II).[5] This figure includes the $2.5 million of the Rockefeller Foundation's support for research in the "great neglected diseases of mankind" and the $2.18 million of the Edna McConnell Clark Foundation's support for schistosomiasis research, but does not include research by pharmaceutical companies in the U.S. This is an impressive beginning, considering that the *total* figure for *worldwide* support for research in all tropical diseases—including parasitic diseases—was estimated at $30 million dollars in the early days of organizing the TDR program (1975). While it will always be difficult to measure the exact degree of influence of the WHO research initiatives on national research policies, the example cited for tropical diseases research in the U.S. indicates that it can be considerable.

## REFERENCES

1. *The medical research programme of the World Health Organization 1958–1963,* World Health Organization, Geneva, 1964.
2. Corning, M. E., 1980, *A Review of the United States Role in International Biomedical Research and Communications (International Health and Foreign Policy),* U.S. Department of Health and Human Services, Public Health Service, National Institutes of Health, National Library of Medicine, NIH Publication No. 80–1638, pp. 196–228.
3. *Sixth Report on the World Health Situation (1973–1977),* 1980, World Health Organization, Geneva, p. 206.
4. Lambo, T. A., 1980, The evolution of WHO's special programmes for research and training, *in: Health Policies in Developing Countries,* Royal Society of Medicine International Congress and Symposium Series No. 24 (C. Wood and Y. Rue, eds.), Royal Society of Medicine and Academic Press, London, Grune & Stratton, New York (U.S. edition), p. 44.
5. Cook, J. E., 1981, Sources of funding for training and research in parasitology, *in: The Current Status and Future of Parasitology* (K. Warren and E. Purcell, eds.), Josiah Macy, Jr., Foundation, New York, pp. 121–135.
6. Goodman, H. C., 1972, The WHO immunology research and training centers: a mechanism for multilateral cooperation in research training, *in: Training of Research Workers in the Medical Sciences,* published on behalf of the Council for International Organizations of Medical Sciences by the World Health Organization, Geneva.
7. The Work of WHO, 1971, Official Records of the World Health Organization, No. 197, pp. ix–x.

8. The Work of WHO, 1975, Official Records of the World Health Organization, No. 229, p. xii.
9. Godal, T., 1977, Two years with a scientific working group, *in: Challenges in Research on Tropical Diseases,* SAREC Report No. R:1 (G. Sterky, ed.), Swedish Agency for Research Cooperation with Developing Countries, pp. 41–47.
10. Lucas, A. O., 1978, Research and training in tropical diseases, *Interdiscip. Sci. Rev.* **3**:220.
11. Knowles, J. H., 1976, American medicine and world health, *Ann. Intern. Med.* **84**:483.
12. In *World Hunger, Health and Refugee Problems,* February 18, 1976, Part VII: International Health and Guatemala Earthquake, Joint Hearings before the Subcommittee on Health of the Committee on Labor and Public Welfare, 94th Cong., 2d sess., U.S. Government Printing Office, Washington, D.C.
13. Message from the President of the United States to the 30th World Health Assembly (May 5, 1977), presented on May 10, 1977, Official Records of the World Health Organization, No. 241, pp. 222–223.
14. The White House, Statement by the President, May 3, 1978, Office of the White House Press Secretary, Washington, D.C.
15. Remarks of Joseph A. Califano, Jr., Secretary of DHEW, May 9, 1978, Press Release, International Communications Agency, U.S. Mission, Geneva, Switzerland.
16. *Congressional Record,* Senate, March 22, 1978, S.4374, U.S. Government Printing Office, Washington, D.C.
17. Norman, C., 1981, AID science funds emerge in new guise, *Science* **213**:626 (August 7, 1981).

# 9

# Histocompatibility Workshops
## Economic Impact

JEAN DAUSSET and FELIX T. RAPAPORT

Few young people today can imagine the conditions under which research was done in the years following the Second World War—the lack of facilities and of scientific contact, and the somewhat limited horizons of the times. The computer had not been invented, there were no plastic tubes or trays, and there was no photocopier down the hall. The immunohematology laboratory where one of us (J.D.) began working was in the basement of the Regional Transfusion Center in Paris, and his bench was an ordinary kitchen table. The work was oriented toward the serology of red blood cells—only the ABO, Rhesus, and MN groups were generally recognized for transfusion matching, and the major research theme of the period was concerned with hemolytic diseases in the newborn. Exchange transfusion had only recently been introduced, and was being practiced daily, not only in newborn babies but also in adults in the fight against anuria following septicemia perfringens, which resulted from voluntary abortion in the absence of antibiotics. This was his first contact with renal failure, which was later to become one of our major preoccupations. Leukocytes and platelets were not even mentioned. However, they soon made an explosive intrusion into the life of the laboratory when he began looking for autoantibodies in the blood of leukopenic

JEAN DAUSSET • Université de Paris VII, Institut de Recherches sur les Maladies du Sang, Hôpital Saint-Louis, 75475 Paris, France. FELIX T. RAPAPORT • Transplantation Service, Department of Surgery, State University of New York at Stony Brook, Stony Brook, New York 11794.

and thrombopenic patients, and observed the first leukoagglutination—after mixing the serum of a polytransfused aplastic female patient with leukocytes obtained from the bone marrow of another individual. Huge agglutinations appeared, after use of the simple technique for determination of ABO blood groups on glass slides. Soon he realized that this reaction could not involve autoantibodies, as the leukocytes of the patient, and a certain number of varieties of leukocytes from others, were not agglutinated. The most likely hypothesis was the presence of isoantibodies, or as they are now called, alloantibodies. The same results were obtained with thromboagglutination; alloantibodies were not "natural," as tests on more than 1000 sera taken systematically from normal nontransfused male donors were uniformly negative.

Having thus established the role of preimmunization by transfusion, the way was clear to pursue studies in immunogenetics. It was necessary to define the leukocyte groups responsible for this immunization through population and family studies. The complexity of the antibodies found in some 50% of the polytransfused patients was evident from the start, pointing to the subsequent polymorphism of the system.

The first panel of reference cells was taken from volunteers from the staff of the National Transfusion Center. The only available technique was leukoagglutination on a glass slide, using defibrinated blood (a technique introduced to eliminate the problem of platelets) [I sometimes wonder how many Erlenmayer flasks I must have shaken for 10 minutes at a time to prepare this defibrinated blood. Sometimes the blood was taken from donors outside the laboratory, often in the family home, and to save time I would continue to shake the flask with one hand while driving the car with the other!] The separation of leukocytes and red blood cells was completed by sedimentation of erythrocytes in blood diluted with dextran. We were a long way indeed from today's easy Ficoll–Isopaque technique.

Soon huge charts covered with columns and crosses began to decorate the laboratory walls, and from these, the analysis (at that time purely visual) was done—these were the days before sophisticated statistical methods. The first leukocyte group, MAC (the initials of the three members of the panel whose leucocytes very frequently remained nonagglutinated), was described thanks to an analogy (admittedly approximative) with only six immune sera. One of the sera had been developed deliberately in a patient, by regular transfusion of the blood of one donor only. Luck was with us—the donor was (as we now know) HLA-A2 positive, and the patient was HLA-A2 negative.

It is not generally understood that at the beginning the roads leading to the unraveling of histocompatibility in the mouse and in man were completely independent. Hemagglutination was the basic technique used in the mouse and leukoagglutination in man. Despite this essential difference it seemed from the start that the two techniques could lead to similar discoveries, and so the analogy was made, notwithstanding the fact that no actual contact between the different research workers in mouse and man had yet been made. Not long afterward the technique of agglutination of human leukocytes was adopted by van Rood in Holland and Payne in the United States. The first discoveries came from these three laboratories. In *Clinical and Biological Immunohematology,*[1] published in 1956, a section just ten lines long then foresaw the occurrence of antileukocyte immunization as a consequence of pregnancy. This possibility was quickly demonstrated by van Rood and Payne independently, thereby uncovering a source of antibodies which were more pure (or more precisely, less impure) than those from polytransfused donors. The research from then on became much more sophisticated—the analysis of thousands of reactions being made easier thanks to one of the first applications of computer techniques to the problem, which permitted van Rood to discover antigens $4^a$ and $4^b$.

Even after modification by van Loghem, who added anticoagulants, the technique of leukoagglutination remained clumsy and poorly reproducible, with much time and effort wasted due to the lack of precision of this devilish technique. An inordinate amount of energy was then spent (uselessly) on the development of another technique—antiglobulin consumption. One recalls at this point that the results of the First Workshop, organized by Amos in 1965 in Durham, were so violently discordant that they were never even made public, and were in fact destroyed. Eventually two techniques still used generally today were developed—lymphocytotoxicity and complement fixation on platelets, introduced by Terasaki and J. and M. Colombani, respectively; these methods have allowed rigorous analysis. Soon the first alleles were described by Payne and Bodmer, and the concept of a complex but unique genetic system called Hu-1, analogous with the H-2 system, was proposed by one of us (J.D.).

The question remained—was there really only one complex system, or two independent loci (the first[2] or leukocyte antigen[3] and the second[2] of the FOUR[4] series)? We remember all too well the 1967 Workshop in Turin, organized by Ceppellini—everyone brought with him technicians and heavy equipment at huge expense, in order to ensure the results of his own work (a refrigerated centrifuge was actually flown over from the States, but could not be released from Customs in time for the Workshop!). All this effort was organized in order to type the blood from some 10 families in "blind" fashion. The results were clear-cut, however—the system was unique. The existence of two allelic series within the system was not really accepted until the following year, and even then not without ambiguity, as shown by the nomenclature adopted in 1968. The numbering of different antigens thus jumped from one series to another (e.g., HLA-A2 and -A5, as one wrote it then, or HLA-A2 and -B5, as it stands now). If the nomenclature committee had met *after* (and not a few hours before) the lectures given by Terasaki, Kissmeyer-Nielsen, and Dausset at the New York Transplantation Congress in 1968, it might have been otherwise, and we might have been able to use a logical listing system today.

It was at this point, and not until this point, that histocompatibility in the mouse and in man finally came together. It is not generally known how much the study of man contributed to study of the mouse. In fact the use of lymphocytotoxicity introduced by Demant at Bar Harbor in 1970 led Snell to change drastically the H-2 map by adopting a bigenic concept for the mouse. D and K became the equivalents of the two human series A and B. The existence of man of many determinants on the same HLA molecule (many "antigenic factors") was not accepted until after description of "public" and "private" antigens in the mouse—a description which was a logical consequence of acceptance of the concept of two loci. However, over the years man's debt to the mouse has become ever greater and the balance is now clearly in favor of the mouse's contributions.

A good example of the length of time required for scientific evidence to be assimilated as dogma is provided by the long-term reticence of many surgeons to admit that histocompatibility could play a role in transplantation. It took nearly 20 years to get this idea accepted. This hypothesis was actually put forth by Dausset in 1958 in a paper describing the antigen MAC.[5] The first experimental proof was given in 1965 by a series of skin grafts presented simultaneously at the Leyden Workshop by the authors of this chapter and by van Rood. In the years that followed, the list of known HLA antigens grew longer and the arguments stronger and stronger. This may be an opportune time also to note the amazing devotion of all the individual and family donors who allowed us to perform hundreds of skin grafts with or without preimmunization. Between these volunteers—parents, children, and other individuals—and the authors there arose an extraordinary degree of confidence and enthusiasm. It seemed that all understood intuitively the stakes for which we were playing and the significance of what we were doing together. There were no limits to what they were prepared to let us do to them (one of them arrived one day with a declaration to this effect signed by his wife!). Now after 20 years many are still linked to our laboratory, and are still ready to help man. They are to be thanked on behalf of the whole scientific community.

It may also be relevant to note at this point that the basic laws of transplantation as they exist today could not have been elaborated without the devotion of these individuals. Indeed the entire serological structure of Hu-1, later known as HLA, would have remained in the realm of theoretical speculation without the indispensable data provided by correlation of these data with actual skin graft survival in human subjects. Much of this effort had its birth in the laboratories of the late John Marquis Converse at New York University, and one of us (F.T.R.) was uniquely privileged to be given the opportunity, almost 30 years ago, to define the survival of skin allografts in man, and to provide the initial demonstration, through skin-grafting experiments, of the actual sharing of transplantation antigens in unrelated randomly selected human subjects. A fortuitous encounter with Dr. Converse in Padua by one of us (J.D.) almost 20 years ago led to the establishment of a long-term intimate collaborative effort between the New York and the Paris laboratories which continues to this day. The rest is history.

The generosity of our volunteers was still not enough to break down the barriers of disbelief among many transplant surgeons, who with rare exceptions were only prepared to accept data they had established themselves. Today the vast statistical evidence finally cancels all doubt. Already the early results seem obsolete when compared with the astonishingly favorable results obtained when HLA-DR locus compatibility is associated with HLA-A or -B. Maybe the surgeons were right after all not to be content with earlier, less spectacular results. It must be said, too, that without pressure from this forceful source, progress would have been a great deal slower, and perhaps we would never have reached the present state of grace. The eight Histocompatibility Workshops constitute a unique example of international stimulation and cooperation, without any precedent in the history of biology. When one looks back at these Workshops, held in the space of 15 years, one realizes that they might never have taken place but for the burning need to prove a point. Skepticism (or blind incomprehension) can have beneficial effects indeed.

It is somewhat amusing to compare the attitude of clinical surgeons confronted with histocompatibility in transplantation with that of internists confronted with associations between HLA and disease. In contrast to the former, the internists only asked to be given proof. Certainly it was an event of major importance to localize on the HLA complex genes implicated in disease. The practical implications of this discovery are still emerging, however, and will continue to emerge very slowly.

It is clear that the role of the HLA complex (and its related genes) is of immense consequence in the immune response. The discovery of genetic control by the major histocompatibility complex of many stages of the immune response is without doubt the crowning achievement of this collective scientific adventure. An increasingly precise knowledge of small segments of the human genome, which intervene decisively in subtle immune mechanisms by recognizing self or nonself and allow the cooperation of the various cell populations involved, is a milestone in biology.

Every new discovery thus constitutes a liberation for mankind, and one can only wish that, if the present rate of discoveries continues, it could bring about control or even eradication of certain diseases. We hope and believe that all young men and women who participate so enthusiastically in this patient analysis are aware that in doing so they are making an essential contribution to the survival of the human species.

The eventual and total beneficial economic impact in unraveling histocompatibility differences between individuals, diseases that are linked to or influenced by the HLA, and the ways the MHC influences and interacts with the immune response is so potentially enormous that it staggers the imagination. One can envision a time in the not too distant future when just knowing the parental HLA genotypes will provide working probabilities of progeny susceptibility to major diseases. One can envision a time when transplantation of almost any organ could be accomplished at will. It can be anticipated that an individual's immune response will one day be modulated selectively to avoid the adverse sequalae of autoimmunity and neoplasia, while simultaneously leaving unaltered the immunities against the myriad of other agents with which we come into contact. Finally, one can envision the approach of an era when a crucial minuscule portion of a genome can be inserted into a deficient one with a resulting restoration of normal immune function. A price tag cannot be placed on these accomplishments any more than on the love of a parent for a child.

## REFERENCES

1. Dausset, J., 1956, *Immuno-Hématologie Biologique et Clinique*, Flammarion, Paris.
2. Dausset, J., Colombani, J., Legrand, L., and Feingold, N., 1968, Le deuxième sublocus du système HL-A, *Nouv. Rev. Franç. Hématol.* **8**:864–868.
3. Payne, R., Trip, M., Weigle, J., Bodmer, W., and Bodmer, J., 1964, A new leukocyte iso-antigen system in man, *Cold Spring Harbor Symp. Quant. Biol.* **29**:285–295.
4. Van Rood, J. J., and Van Leeuwen, A., 1963, Leukocyte grouping: A method and its applications, *J. Clin. Invest.* **42**:1382–1390.
5. Dausset, J., 1958, Iso-leuco-anticorps, *Acta Haematol.* **20**:156–166.

# 10

# A Model for the Funding of Clinical Faculty in a Municipal Hospital

CHESTER W. SCHMIDT, Jr., PHILIP D. ZIEVE, and BURTON C. D'LUGOFF

## INTRODUCTION

During the decade of 1969–1979 medical school revenues increased from $1.4 billion to $4.9 billion. Federal support for biomedical research doubled over the same period of time, but fell from 30% of total revenues in 1969 to 18% of total revenues in 1979. While the percentage of federal research dollars was decreasing, income from academic practice plans (professional fees) rose from 4% of revenues in 1969 to 14.8% in 1979.[1]

Clearly academic practice plans have become increasingly important as a source of financial support for the faculty of academic centers.[2] Despite the vexing problems that clinical activities create for academic medical centers—integration of teaching and research with "private" modes of practice, academic advancement of clinical faculty, faculty sharing of financial resources—it is clear that practice plans can support the teaching and research goals of academic faculties.

This chapter describes the planning, implementation, and development of one such plan in a municipal hospital, a major affiliate of a medical school.

CHESTER W. SCHMIDT, Jr., PHILIP D. ZIEVE, and BURTON C. D'LUGOFF • Departments of Medicine and Psychiatry, Baltimore City Hospital and Johns Hopkins University School of Medicine, Baltimore, Maryland 21224. The text of this chapter was first published in the *New England Journal of Medicine*, January, 1981, and it is reprinted here with revisions.

## THE SETTING

Baltimore City Hospital consists of a 400-bed general acute-care hospital and a 200-bed chronic-care hospital. There were 8218 admissions to the acute-care hospital during the fiscal year 1979–1980, accounting for 103,698 hospital days. The combined outpatient clinics had 65,818 visits during the same year, and there were 35,295 visits to the emergency room.

The facility is in an unusual location for a municipal hospital: it is near the edge of the city in a lower-middle-class neighborhood that houses primarily the families of blue-collar workers employed in nearby industries. Consequently, in contrast to patients in many other municipal hospitals, a substantial number of these patients have commercial insurance, and most of them have some kind of third-party medical coverage.

The staff during the fiscal year 1979–1980 consisted of 139 full-time and 97 part-time professionals. There were 131 physicians, two dentists, and six psychologists with doctoral degrees. The professionals were distributed in nine departments: anesthesia, medicine, neurology, obstetrics and gynecology, pathology, psychiatry, radiology, dermatology, and surgery.

Of the full-time physicians, 101 are hospital-based specialists and 32 are primary-care physicians. The primary-care physicians have been trained in internal medicine or in pediatrics. Overall, 85% of the full-time staff are board certified.

All members of the full-time medical staff are full-time faculty of the John Hopkins University School of Medicine. At any given time 25% of the medical students who are on clinical rotations are based at City Hospital. Required clerkships are provided in surgery, pediatrics, psychiatry, and medicine.

There are shared house-staff programs with the Johns Hopkins Hospital in the departments of surgery, obstetrics and gynecology, anesthesiology, neurology, radiology, pathology, dermatology, and psychiatry. The departments of medicine and pediatrics have independent residency programs. Residents in all programs hold simultaneous appointments as fellows in the university. The total number of house-staff officers and fellows is 118.

The research activities of the hospital can be measured by the fact that $2.9 million in research grants and contracts was awarded to the faculty during the 1979–1980 fiscal year. Some 7% of the faculty members are involved in full-time research and another 25% are funded on a fractional basis for various amounts of research activity. Approximately 60% of the amount granted supports basic laboratory research and the remainder supports clinical research.

## FORMATION OF THE CPPA

During the late 1960s Baltimore City Hospital was in a precarious position. Because of the fiscal constraints that chronically affect the city, it was extremely difficult to obtain approval to fund new positions, to expand existing programs, or to establish new ones. However, the major threat to the institution was the progressively declining clientele, which was attributable to the dispersion of that portion of the patient population, now supported by Medicaid and Medicare, who had been forced to accept Baltimore City Hospital as their only available provider of care until the advent of these programs. To compound the problems, the salaries and fringe benefits of the medical staff, paid by the city of Baltimore and the Johns Hopkins Medical School through Baltimore City Hospital, were too low to be competitive with those paid by other teaching hospitals in the region.

In response to these threats to the survival of the hospital, the medical staff obtained the permission of the city of Baltimore to incorporate as an independent organization in October 1972.[3] The feasibility study and the accounting and legal work that were necessary to create the organization were supported by an education fund of the hospital, which had been the recipient of that part of the per-diem payment by third-party carriers that was allocated to physicians' services. The initial contract with the city stipulated that the medical staff, incorporated as Chesapeake Physicians Professional Association (CPPA), a nonprofit corporation, would deliver all the inpatient and outpatient care at the hospital, provide all necessary educational supervision, and continue to administer the service and academic programs. During the first 9 months of operation the city reimbursed the corporation $1.6 million for these activities. Beginning on July 1, 1973, and for each fiscal year since then, the city has continued to provide that appropriation, so that the $1.6 million has been paid on the basis of 12 months of activity rather than the initial 9 months. Since the city bore a cost of $2.2 million in professional salaries for 77 full-time staff members before the inception of CPPA, the taxpayers have saved $0.6 million per year. There has been no increase charged to the city because of inflation or because of expansion of the staff, which now comprises 139 full-time workers.

## GOVERNANCE OF THE CPPA

The CPPA is a physician-owned, physician-governed, nonprofit corporation. Ownership by the 118 physician members is created through the issuance of one share of nontransferable stock, with a par value of $1.00, to each member. If the corporation is dissolved for any reason, all assets will go to Baltimore City Hospital rather than to the members.

The ultimate authority of the corporation is vested in the board of directors, which is composed of all chiefs or departments, three members elected at large, the administrative director, and the controller of the corporation. The board of directors elects the president, two vice-presidents, and the secretary-treasurer. The president has the day-to-day responsibility of managing the executive and finance committees of the board. Stockholders meet quarterly to receive committee reports, to approve program development, and to comment on the progress of the corporation.

The corporation is dependent on the relatively few physician members who are interested in overseeing a large business organization and are able to do so. The president of the corporation now spends two-thirds of his working time in the administration of the CPPA. Since he is also chief of the department of psychiatry, that duty requires substantial delegation of authority to senior physicans in his department. We have considered hiring a full-time physician-administrator, but we have not yet done so.

Governance of the corporation at the department level is complicated by the fact that each department chairman must report to three corporate entities. For the CPPA, the department chairman is the budget manager responsible for fiscal affairs, reporting to the board of directors of the corporation; for the hospital, he is the clinical chief, reporting to the executive director of the hospital, the medical board, and the hospital commission; and for the Johns Hopkins University School of Medicine, he is academic director of the department, reporting to his department chairman at the medical school. Despite the fact that he must answer to multiple authorities, the department chairman at the Baltimore City Hospital is given autonomy in conducting the affairs of his department. The only real constraints on his authority are fiscal and academic. All salaries must be approved by the hospital commission and the dean of the medical school, and must be consistent with the resources of the CPPA; in addition the fees charged to patients must be approved by the hospital commission. In the academic realm, it is required that appointments and promotions move through the channels of the parent department of the medical school.

The intent in the establishment of the CPPA and the driving force behind it have been academic, but the plan has necessarily required a major commitment to fiscal responsibility. Because academic physicians usually have no expertise (and in many cases, no interest) in fiscal affairs, the establishment and operation of this kind of corporation require the participation of a highly qualified administrative staff to ensure the proper management and expansion of the program. Business and financial expertise in the CPPA is supplied by the administrative director, who is a certified public accountant, and by his staff. The cost of that operation for 1979–1980, including relatively expensive continuing legal and accounting fees, was $2.4 million, or 16% of the total expenses of the corporation during that year.

## MODE OF PRACTICE

The policy of the corporation dictates that each patient admitted to the hospital be assigned to a member of the faculty who performs an admission history and physical examination, sees the patient at least 6 days a week, and is identified to the patient as the responsible physician. To bill for a surgical procedure, the faculty member must be in attendance when the procedure is performed. The corporation has accepted assignment of fees by the local third-party carriers, all of which have agreed to the mode of practice. Initially the carriers allowed countersignature of house-staff notes as evidence that the attending physician was present daily. In the past year they have insisted that the faculty member write progress notes on the days when charges are rendered unless there is other evidence in the chart (nurse's notes, for example) that the attending physician has seen the patient on that day.

## GROWTH OF THE CPPA

The number of employees of the CCPA and the revenue and expenses of the corporation in 1972–1973 and in 1979–1980 are shown in Table I. The growth in revenue has come from several sources. Patient fees have increased because of increased productivity and an improvement in the collection ratio. The overall collection ratio is computed as a weighted average of the collection ratios for each component by the third-party patient mix. The sources of noncontractual revenue from patient care and the collection ratios from these sources in 1972–1973 and 1979–1980 are shown in Table II. It is important to

**Table I.** Growth of the Chesapeake Physicians Professional Association (CPPA) over 8 Years

| Category | Fiscal year 1972–1973 | Fiscal year 1979–1980 | | Growth |
|---|---|---|---|---|
| | | Actual | In 1972 $ | |
| Revenues ($) | 2,688,007 | 14,202,876 | 7,934,567 | 5,246,560 |
| Expenses ($) | 2,505,465 | 14,134,317 | 7,896,266 | 5,390,801 |
| No. of full-time professionals | 77 | 139 | — | 62 |
| No. of nonprofessional employees | 43 | 300 | — | 257 |

**Table II.** Sources of Patient-Care Revenue

| Source | Fiscal year 1972–1973 Percent of total | Fiscal year 1972–1973 Collection ratio | Fiscal year 1979–1980 Percent of total | Fiscal year 1979–1980 Collection ratio |
|---|---|---|---|---|
| Medicare | 14 | 0.60 | 32 | 0.71 |
| Medicaid | 21 | 0.33 | 19 | 0.40 |
| Blue Shield | 15 | 0.65 | 19 | 0.89 |
| Commercial insurance[a] | 5 | — | 10 | — |
| Self-paying | 45 | 0.20 | 20 | 0.30 |
| Total | 100 | 0.37 | 100 | 0.58 |

[a]The collection ratio is included in the self-paying category.

note that despite the substantial increase in collection of net fees for services rendered, the corporation still loses 42¢ of every dollar charged to its patients. Allowance for free or charity service has had profound financial effects on the corporation and will continue to have such effects.

The salaries of the physician members of the corporation are regulated by the department chairmen at Baltimore City Hospital and are subject to approval by the hospital commission (the board of trustees of the hospital) and by the dean of the medical school. All faculty members are salaried on a strict full-time basis. Although faculty members may generate income from multiple sources, each member receives one check from the CPPA. The contract between the CPPA and each member permits no clinical activity outside the CPPA. Royalties and honoraria may be retained by faculty members. There are no salary incentives and no provisions for augmenting salaries through extra practice.

The salaries have risen from an average of $22,000 per professional in 1972 to an average of $49,000 at present ($49,000 is the equivalent of $27,000 in terms of 1972 dollars, a 22% real increase). Considerable salary differences exist between departments, reflecting market pressures for recruitment and retention of academic faculty. For example, the average salary in the department of surgery (the highest paid department on the average) is $65,400, whereas the average salary in the department of pediatrics (the lowest paid department) is $40,322. The Johns Hopkins Medical School, the parent academic institution, has set the goal for its academic salaries to be at the 80th percentile of the ranges by academic rank that are published by the Association of American Medical Colleges (AAMC) for Middle Atlantic Medical Schools.[4] The CPPA has been required by the dean's office to follow suit. The differences between departments are thus essentially the differences reported on a yearly basis by the AAMC.

Salaries and fringe benefits to the physicians represent 68% of all expenses. Excess revenue over expenses has averaged 3% of net income per year, so that up to the present few dollars have been available for direct funding of research by the CPPA. The CPPA, however, had indirectly supported research and teaching activities through the expenditure of developmental funds for new services and for equipment used in clinical research and teaching, totaling approximately $1.2 million during the past 8 years.

Some departments are producers of net revenue; others incur a net cost to the CPPA. Since all funds are pooled, the departments with deficits are subsidized by the departments that have an excess of revenue over expenses. Within the past year the CPPA has instituted a departmental research-incentive program that returns a minimum of 60% of departmental excess revenue to the departments for education and research, provided that all corporate financial obligations and goals have been met. The organization of the CPPA as an administratively centralized, multispecialty group practice is unique within the Johns Hopkins medical institutions. Other practice groups within the medical school are organized along departmental lines, with some departments sharing centralized billing functions. The reason that the CPPA was organized in a different fashion is that certain departments (anesthesia, psychiatry, pediatrics, and pathology) would not have been financially viable when the corporation was established because there was no endowment money available to make up their deficits. The entire faculty decided that an "all for one and one for all" strategy had the best chance of succeeding. Two of these departments, pediatrics and psychiatry, still require subsidization. However, the magnitude of the departmental deficits has been substantially reduced, from more than $250,000 to less than $75,000.

## OUTREACH PROGRAMS

To stabilize and improve the patient base of the hospital in keeping with the original intent of the corporation and the hospital, the CPPA has established a number of outreach centers during the past 7 years. There are now seven of these programs, all located within the geographic area served by the hospital and staffed by full-time members of the corporation. These centers range from traditional fee-for-service practices to prepaid plans, including a federally qualified health-maintenance organization (HMO), a state-sponsored HMO, and a Blue Cross-sponsored HMO. These outreach centers now employ 32 physicians, generate $8 million in revenue (figures included in Table I), and account for approximately 2000 admissions and 16,000 patient days each year. The programs are still expanding, and it is estimated that by the end of 1980 the total patient population served by these centers will be 63,000.

Most of the faculty members attracted to these programs have joined the CPPA immediately after completing their house-staff training. They are obligated to spend 24–32 hours per week delivering care in the ambulatory programs and to spend the remainder of their time at Baltimore City Hospital, where they are involved in teaching, clinical activities, and research. Of these practioners 25% have subspecialty training and use their hospital time to pursue subspecialty interests as consultants, teachers, and research scientists. Funding of their entire salaries comes from their ambulatory activities.

These programs are used for elective medical student teaching and for required house-staff training. To dat only a few medical students (five per year) have completed clinical electives in these programs. However, the medical school has indicated that it does not intend to replicate these types of programs within the Johns Hopkins medical institutions, so that these ambulatory programs will be a teaching resource for students interested in primary care.

House-staff training in ambulatory medicine is another matter. Baltimore City Hospital and the Johns Hopkins University School of Medicine have received a grant from the National Institutes of Health that provides a track for internal medicine and pediatric training in the ambulatory programs. In addition some departments (dermatology, for example) already conduct at least half their outpatient training of house staff in the CPPA ambulatory sites.

## ANCILLARY ACTIVITIES

### MALPRACTICE INSURANCE

In 1976, because of escalating malpractice premiums and the possibility that it might be unable to purchase any commercial malpractice insurance policies at all, the CPPA established a captive insurance company in Colorado, the Chesapeake Casualty Insurance Company (CCIC), which now insures the first $650,000 of liability for each member of the corporation. A commercial insurance company provides the coverage from $650,000 to $1 million per insured worker per year, and provides $2 million per incident and $5 million in the aggregate per year above that covered by the captive company. The city of Baltimore has participated from the beginning of this venture, indemnifying the house staff and therefore paying the premiums for their coverage by CCIC and by the commercial carrier. For initial capitalization it was necessary for the CPPA to obtain a letter of credit for $1.5 million to establish the insurance company, and to obtain the letter the corporation pledged retirement funds that the physicians and corporation had contributed to an annuity. The experience to date has been good: the malpractice premiums to the company currently show assets of $1.6 million, and in fiscal 1979, excess revenues of $119,000.

### RESEARCH AND EDUCATION

In 1976 the CPPA established a separately incorporated fund for education and research, the Chesapeake Education and Research Trust. This foundation is the repository for grants that are not funded through the medical school or the city of Baltimore, and for donations to the corporation or to one or more if its members. Although the trust has not been successful to date in attracting large amounts of unrestricted money, it has served as a convenient repository for the research funds of individual physicians.

## ACCEPTANCE OF THE CPPA

The acceptance of the CPPA by the various groups involved with it has varied somewhat but has been generally favorable. The medical school has been impressed with the organization's ability to establish new clinical and academic programs in the hospital and in the community. The medical school has also recognized the important role of the financial growth of the CPPA in stabilizing Baltimore City Hospital, thereby preserving needed teaching sites for medical students and house officers. Finally, the school has come to appreciate that the CPPA now funds the salaries and benefits for 20% of its full-time clinical faculty.

The city of Baltimore initially looked on the CPPA with some suspicion despite the fact that it had allowed the organization to form, and a continuing educational effort has been required to convince many people in the municipal government that the individual physicians have not had as their primary intent the accumulation of wealth through the establishment of this practice plan. With the active support of the hospital administration and of the hospital commission, both of which recognize the value of the corporation, these suspicions have been largely, but not completely, allayed.

The physicians in the group were forced to adjust to a dramatic change in daily routine as the result of the change in the mode of practice at Baltimore City Hospital. Attending physicians in general medical services, for example, increased their clinical time from 3 hours a week to approximately 16 or more. Similar increases occurred in other departments. In addition, physicians who perform procedures (surgeons, obstetricians and gynecologists, cardiologists, and others) were obligated to devote much more of their time to clinical care than they had previously given. Especially with respect to the surgical specialists, the difference between private and academic practice became less perceptible, and the predictable question has been raised of whether such intense involvement in clinical work would be better accomplished in a setting that allowed greater remuneration. Nevertheless the attrition rate has averaged only 5% per year, and many new members have been recruited.

Initially the house staff, accustomed to the nearly total responsibility for patients that they had traditionally enjoyed at a municipal hospital, were concerned about the establishment of a clinical practice plan at Baltimore City Hospital. However, because the attending physicians were able to teach more effectively since they had much better knowledge of the patients than they had had before, because the staff was clearly strengthened by the attraction of new and diverse personnel, and because it became clear that there were opportunities for the more talented members of the house staff to remain in the group on completion of their training programs, no major resistance to the CPPA has appeared. So far as it can be determined the CPPA has had no impact, positive or negative, on the recruitment of house staff for the independent residency programs in medicine and pediatrics. In addition, after the graduation of the house staff who were present at the founding of the corporation, the subsequent house staff, who know no other system, did not have the same prejudices that the initial group had had. Although the responsibility of the house staff for patient care has unquestionably diminished somewhat with the advent of the CPPA, it is generally believed by both faculty and house staff that the overall effect of the plan on the training program has been positive.

The patients of Baltimore City Hospital have accepted the assignment of CPPA physicians to their care with hardly a sign of protest. Patients who are referred to the hospital by private physicians are referred back to those physicians on discharge. Each patient enrolled in one of the CPPA ambulatory programs is followed in the hospital by the primary-care physician, who becomes the attending physician on the case. Patients who do not have family doctors are offered the opportunity to enroll in the ambulatory program that suits their needs.

The addition of the full-time attending physician has caused some billing problems for patients, because they receive separate bills from the hospital for its charges and from the CPPA physician for professional services. However, since the plan accepts third-party assignments in all instances and scales down its charges according to the patient's ability to pay if third-party coverage is not available, even the problem of billing has not been considered major.

## IMPACT ON THE ACADEMIC PROGRAMS OF THE MEDICAL SCHOOL

The first thing that must be said about the effect of the CPPA on these academic programs is that it has saved irreplaceable teaching beds and a substantial portion of the full-time teaching faculty for the medical school. The CPPA has developed a series of primary-care programs for ambulatory patients which are available to the school for medical student education and house-staff training and will not be reproduced elsewhere within the Johns Hopkins institutions. The promising future of the academic life of the Baltimore City Hospital campus was a major factor in the recent movement of the clinical laboratory research programs of the National Institute of Drug Abuse from Lexington, Kentucky, to the Baltimore campus.

A less tangible but nevertheless important impact on the academic programs of both Johns Hopkins and Baltimore City Hospital has been the change in the attitude of Hopkins toward the hospital. Both the hospital and faculty have a new respect for each other as partners in the future development of the Baltimore City Hospital campus as an integral part of the institution. The change in attitude is a result of the accomplishments of the faculty, who have used the CPPA to develop the potential of the Baltimore City Hospital's medical campus.

## ISSUES

There are numerous consequences of relying on clinical practice to fund the salaries of academic faculty, with effects on organizations such as the CPPA.

### RELUCTANCE OF PAYERS TO FINANCE MEDICAL EDUCATION

Some believe that medical education can be separated tidily from patient care and that patient fees ought to pay only for the services rendered. Third-party payers, governmental policy analysts, and health planners often question the costs associated with the mixture of service and education in a teaching hospital. There is a relentless pressure to account for every dollar of revenue derived from patient fees, even though that philosophy does not adapt realistically to the processes that bring medical care and medical education together in a hospital setting. Education is conceived by such persons solely as didactic instruction that takes place in a lecture hall or laboratory. Bedside teaching, case conferences, operating room and procedure supervision, and the entire multifaceted complex of exposures of the medical student and house staff to the activities of the attending physicians in a supervised dialectic process—the hallmarks of the Flexnerian revolution in medical education—are more or less foreign to the cost accountant and the medical planner. The constant, dynamic shifting of the roles from service to education and back defies the strictly economic form of analysis. Insofar as the practice of medicine remains a clinical art that is enhanced to a high level of artistry through apprenticeship, this method of acquiring the art cannot be easily discarded.

The costs of apprenticeship training have traditionally been borne by the purchaser of any particular service or profession. This principle is most clearly visible in the construction trades, in which apprentices, journeymen, and master craftsmen all participate in teaching apprenticeships on the job while performing their craft side by side. The cost of time spent by the master craftsmen while they are supervising an apprentice is paid fully, not at a reduced rate, and it is not identified as illegitimate.

There is at least one other practical justification for supporting the apprenticeship system. In a teaching hospital such as Baltimore City Hospital, the total amount of money earned by the faculty's clinical efforts is limited by the acceptance of all patients, whether or not they can pay. The CPPA, as noted above, collected only 58% of its gross charges during the fiscal year 1979–1980; one reason for this figure is that our self-paying patients can only afford to pay 30% of their fees. The 58% collection ratio in effect resulted in a donation of $4,605,155 in free care to the CPPA's patients.

These donated sums eventually affect the total pool out of which the faculty is paid, thereby contributing to the lower salaries of these academic physicians. The CPPA pays its professionals salaries and benefits that are one-third to two-thirds lower than the income that they would earn in private practice in the metropolitan area. Under such circumstances the accusation that faculties are being subsidized by the health-care system is open to question. It appears that such teaching faculties not only care for patients who are usually ignored by the private sector of medicine, but also absorb both the financial losses associated with treatment of that patient population and some portion of the cost of teaching the house staff.

## THE STATUS AND PROMOTION OF CLINICAL FACULTY

Another major issue that has arisen because of the shift in the source of funding of academic faculties is the changing composition of such faculties. Funding during the past 30 years by research grants from the National Institutes of Health has resulted in an academic establishment whose preeminent interest has been bench research and scientific publication. Clinical activities and teaching ability, though not precluded by research interest, have not been prime qualifications for entry into medical faculties, nor have they usually been sufficient criteria for advancement to the professorial rank. In keeping with the proposition that "he who pays the piper calls the tune," clinical activities and teaching ability in themselves will have to be increasingly accepted and recognized for the purposes of promotion and advancement.

The Johns Hopkins Medical School is also grappling with this issue. In all probability persons who demonstrate excellence in clinical care and teaching may have the opportunity to advance to the level of associate professor. One means of addressing this issue that applies particularly to Hopkins, which has only one full-time academic track, is the use of part-time academic appointments for persons who conduct no research. The criteria for academic advancement for part-time faculty already take into account excellence in clinical practice and teaching.

## THE NEED FOR THE FACULTY TO SHARE INCOME

Medical schools and their faculties must recognize that their institutions are large and complex organizations that rely on the diverse talents of many people. The traditional conflicts between basic science and clinical departments must be set aside. All members of the faculty should have an equitable share of the institution's resources.

The substantial differences in salaries between different clinical specialties—principally reflections of third-party reimbursement schedules for procedures and operations—constitute a major problem in this regard. The excess revenue generated by members of the clinical faculty in procedure-oriented specialties must be used to support non-revenue-producing activities that are vital to the academic goals of the medical school and of the teaching hospital. One of the reasons for the success of the CPPA to date has been the willingness of the divisions and departments that generate excess revenue to distribute it according to the needs of the organization as a whole.

## THE COMPETITION BETWEEN ACADEMIC AND PRIVATE CLINICIANS

As academic faculties organize to increase their patient-care activities there are certain settings in which they may find themselves viewed as unfair competition to existing community practitioners—a new extension of the perennial town-versus-gown controversy that has always afflicted universities and their communities. In other settings this competition may not be a problem, for example in public general hospitals of university medical centers located in areas that are almost totally devoid of private practitioners. On the contrary, these practice plans may serve to accomplish the goals of geographic accessibility to care that have long defied the planning skills of social and governmental reformers. Largely because of the scarcity of private practitioners in the service area of the Baltimore City Hospital, the CPPA has served such a need without conflict. In communities where the university medical center as a tertiary-care facility has long enjoyed an established referral network of community-based practitioners, the only effect of a program such as the CPPA may be that the bills will no longer be rendered by the university but by a professional corporation or practice plan instead.

## CONCLUSIONS

Federal support, and to a lesser extent private support, for teaching and research are decreasing at the very time inflation is eroding the financial base of our medical centers. In order to preserve activities which are vital to our institutions and our patients we must not only lobby at all levels of government for increased support, but we must also develop new sources of funding. If nothing else, current experience should teach us the necessity of placing our financial eggs in as many baskets as we can find. Revenue from practice plans is a potential new source of funding for all activities of our medical centers. Properly managed, these plans can improve patient care, enhance the quality of teaching, pay for the clinical activities of the faculty (thereby freeing up other monies for basic science salaries), and directly support research through excess revenues.

What is the hitch? If the practice plans can accomplish all these wonderful things, why isn't everyone doing it? The answer is because there are difficult problems associated with the development and management of these plans within academic settings. Many of those problems have been discussed, but fundamental to most of these issues is the capacity of academic institutions to commit themselves philosophically and pragmatically to straightforward clinical enterprises. Concerns about the distortion of institutional educational and research goals are legitimate, but experience over a 9-year period makes us hopeful that we can continue to use our practice place to further our academic interests. We believe others can do the same.

## REFERENCES


1. Medical education, 1980, *J. Am. Med. Assoc.* **224:**2810–2827.
2. Rogers, D. E., 1980, On preparing academic health centers for the very different 1980s, *J. Med. Educ.* **55:**1–12.
3. Sterioff, S., Jr., and Smith, G. W., 1973, Incorporation of the entire medical staff of a municipal hospital: One solution to the academic problem of the 1970's, *J. Surg. Res.* **14:**105–108.
4. Association of American Medical Colleges, 1980, Report on medical school faculty salaries, 1979–1980, Association of American Medical Colleges, Washington, D.C.


O.P.C. ABERYSTWYTH

# 11

# Economics of American Health Care

VICTOR R. FUCHS

Most physicians in the United States today have never known a time when the real resources at their disposal were not increasing at a rapid rate. In the 1980s, however, the flow of funds for patient care, research, and teaching is likely to slow appreciably because the economy will be growing at a slower rate and because the difference between rates of change of health expenditures and the gross national product (GNP) is likely to narrow. These prospects, along with the almost certain rapid growth in the *number* of physicians for at least another decade, pose a critical financial challenge to American physicians in private practice and in academic medicine.

The past 30 years, during which most physicians who are currently active began their careers, have brought major changes to American medicine. Scientific and technologic advances have greatly strengthened the physician's ability to intervene decisively in the course of disease or injury. Medical organization, traditionally based on solo general practice, has moved substantially in the direction of group and specialty practice. Medical education has been set on a firm foundation of basic science and medical research has exploded into a major activity for thousands of physicians.

Not least in importance has been the dramatic increase in expenditures for health care. Between 1950 and 1979 national health-care expenditures, adjusted for inflation by the GNP deflator, grew at an annual rate of 5.9%. The GNP itself, similarly adjusted for inflation, grew at 3.5% per annum. If one takes a somewhat more limited

VICTOR R. FUCHS • Department of Economics, Stanford University, and National Bureau of Economic Research, Stanford, California 94305. Reprinted by permission of the *New England Journal of Medicine*, 1981, **306**:1487–1490.

measure of health-care expenditures that excludes dental care, nursing homes, and other activities that bear little relation to the work of physicians, the discrepancy in growth rates is even slightly larger. The average gap in growth rate between GNP (the nation's output of goods and services) and "physician-related" expenditures was 2.5% per annum.

The relationship between GNP, health expenditures, and number of physicians can be explored further with the aid of the following identity:

$$\left(\mathrm{GNP} \cdot \frac{\mathrm{EXP}}{\mathrm{GNP}}\right) \div \mathrm{PHYS} \equiv \frac{\mathrm{EXP}}{\mathrm{PHYS}}$$

The statement that health expenditures per physician is equal to the GNP multiplied by health expenditures' share of GNP, all divided by the number of physicians, is of course tautological. However, it can direct attention to relevant variables and help in the formation of judgments about how changes in these variables may affect one another in the future. The rates of change of these variables between any two years are also related tautologically, except that the identity is additive rather than multiplicative because the rates of change are based on the logarithms of the values in the initial and terminal years.

Past trends in these variables are presented in Table I. We see that in the 1929–1950 period the difference in growth rates of health expenditures and GNP was only 1.4% per annum. The number of

**Table I.** Rates of Change of Key Health Economics Variables, Selected Periods 1929–1979 (Percent per Annum)[a]

| Period | $\hat{\mathrm{GNP}}$ + | ($\hat{\mathrm{EXP}} - \hat{\mathrm{GNP}}$) − | $\hat{\mathrm{PHYS}}$ ≡ | ($\hat{\mathrm{EXP}} - \hat{\mathrm{PHYS}}$) |
|---|---|---|---|---|
| 1929–1950 | 2.5 | 1.4 | 1.4 | 2.5 |
| 1950–1965 | 3.7 | 2.3 | 1.8 | 4.2 |
| 1965–1979 | 3.3 | 2.7 | 3.0 | 3.0 |

[a] $\hat{\mathrm{GNP}}$, rate of change of the Gross National Product deflated by the GNP deflator; $\hat{\mathrm{PHYS}}$, rate of change of the number of physicians; $\hat{\mathrm{EXP}}$, rate of change of health care expenditures (excluding dental and "other" professional service, nursing homes, eyeglasses and appliances, construction, and "other" services) deflated by the GNP deflator.

physicians also grew at 1.4%; the expansion of expenditures per physician matched the growth of GNP at 2.5% per annum. The two periods since 1950 were both marked by rapid growth of GNP. Until 1965 this growth was attributable primarily to an unprecedented increase in productivity. Gains in productivity were not as rapid between 1965 and 1979, but an unusual expansion of the labor force helped boost GNP. In both periods the health expenditure–GNP differential was substantial. During 1950–1965 it was fed primarily by the rapid diffusion of private health insurance. After 1965 government programs such as Medicare and Medicaid provided the major fuel for expansion. The rate of change of physicians was modest during 1950–1965, but accelerated sharply after that time; the rate of growth of expenditures per physician decelerated after 1965 but remained at a substantial 3.0% per annum.

What does the future hold? Clearly it is dangerous to infer causality from an identity, and it is always hazardous to make forecasts about variables that have shown great variation in the past. Nevertheless I believe it is possible to provide some informed judgment about what is likely to happen during the 1980s. First, it seems unlikely that the GNP will grow as rapidly as it has since 1950. The first 20 years after World War II were unusual ones for productivity gains, and no one expects such gains to be repeated in the coming decade. Furthermore the principal forces contributing to rapid labor force growth since 1965—the baby boom of the 1950s and an upsurge in female labor force participation—will both be attenuated in the 1980s. However, without unusual gains in productivity or size of the labor force, a GNP growth rate of 3% is about all that can be expected.

Second, it seems extremely likely that the number of physicians will continue to grow at a rapid rate. An increase of 3% per annum is probably a conservative estimate, judging from current medical school enrollments. If the number of physicians grows as rapidly as the GNP, it follows from the identity that the growth of expenditures per physician will equal the "gap," i.e., the rate at which change in health expenditures exceeds the change in the GNP.

Because the direction of causality is not a settled matter, one could argue that expenditures per physician will continue to grow at 3% or 4% per annum, and that therefore the health sector's share of

GNP must expand equally rapidly. Dramatic technologic advances in medicine that are both very effective and very expensive might produce this result. It seems to me, however, that the reverse scenario is more likely. I believe that the forces opposing the continued expansion of the health sector are so strong that the "gap" will probably not exceed 2% per annum. If that happens, expenditures per physician will grow at an unprecedentedly low rate.

What are these opposing forces? First, the government's contribution to health expenditures (presently almost half the total) will be restrained by opposition to higher taxes in general and by increasing competition from other claimants for government funds, e.g., defense, energy, and control of crime. Second, there will be increasing resistance in the private sector to ever-higher health insurance premiums. Employer groups are already forming in many local areas specifically to deal with health-care costs. One simple bit of arithmetic explains why resistance to the expansion of health-care expenditures is likely to grow on both the public and private fronts. When the health sector represented only 5% of the GNP, an excess growth of health expenditures over GNP of 2% per annum implied a gain for the health sector of 1 percentage point (i.e., to 6%) in a decade. Now that the health sector is close to 10% of the GNP, an identical "gap" of 2% per annum implies that the health sector captures an additional 2 percentage points of the GNP in the course of a decade. Because so much more is at stake now, the resistance on the part of those who must forgo alternative uses of funds is likely to increase.

How will this slowing down in the rate of growth of expenditures per physician be achieved? One likely prospect is for a decrease in the average annual hours worked per physician. Indeed I believe such a decrease is already occurring to a greater extent than is revealed in most surveys of physicians' time input. Much of this decrease takes the form of days off, long weekends, longer vacations, and sabbaticals rather than a reduction in hours per day. A direct-observation survey of a population of general surgeons found that the modal day of work was 9 hours, but the average annual input per surgeon was under 1800 hours.[1] To the extent that the average physician will be working fewer hours per year in 1990 than in 1979, he or she is likely to be hospitalizing fewer patients, writing fewer prescriptions, and in general generating lower expenditures.

The principal vehicle of deceleration, however, is likely to take the form of changes in the organization and financing of medical practice. Expenditures per physician, it must be remembered, constitute the gross revenues of the physician, the hospitals, the drug companies, and other providers of medical goods and services. The physician's own revenues account for only a minor part of the total, but the physician's decisions have great influence on the size of that total. Thus physicians, in their own self-interest, are likely to move toward modes of practice that hold down that portion of the expenditure package that finances hospitalization, drugs, and the like, rather than allow their own incomes to be reduced.

Many such modes of practice are already in place—prepaid group practice, independent practice associations, primary-care networks—and still others will be created in the years ahead. The central feature of all such modes is that the physician's own income is explicitly or implicitly tied to the total cost of care. When there is a fixed amount of money to be divided, the greater the expenditures for hospitals and drugs the less is available for physician reimbursement, and vice versa. Physicians practicing in such modes are moved to reexamine their approaches to diagnosis and treatment in order to provide maximum care for their patients per dollar expended. Most present third-party reimbursement systems do not offer that incentive to individual physicians or patients.

To be sure, there are potential problems with such modes, including the possibility that patients will receive too little care. Moreover, the great majority of practicing physicians are comfortable with existing fee-for-service systems and will not eagerly accept such organizational changes. The logic and the force of external circumstances, however, is likely to prevail. As Alain Enthoven concluded in his discussion of the independent practice association in Minneapolis, "If economic survival in the face of competition is at stake, physicians will impose on themselves controls they would never dream of accepting if imposed by government."[2]

If some curtailment in the rate of growth of health expenditures is inevitable, these physician-centered modes of control are clearly preferable to the only feasible alternative—centralized regulation. These modes permit much greater flexibility in adjusting to the needs and preferences of patients at the local level where the care is actually provided. Most important, they keep medical decision-making where it belongs—in the hands of the physicians. Thus those practicing physicians who have a clear view of the long-run interests of their patients and themselves are likely to move toward these new forms of organization positively and constructively.

The trends described in this chapter not only will have an impact on physicians in private practice, but also present a major challenge to medical school faculties. Indeed because the expansion of medical schools has been so rapid since 1950, and because the slowing down will be so marked, the coming decade is likely to be particularly traumatic for academic physicians in patient care, research, and teaching.

Revenues from patient care currently provide a significant portion of faculty incomes at most if not all U.S. medical schools. In the future these revenues are likely to be subject to the same competitive pressures that will face physicians in private practice. Moreover the solution of reducing hospital utilization in order to protect physician income will be especially controversial in universities that bear fiscal responsibility for the hospital as well as the medical school.

The new, tight financial environment is also likely to result in some change in emphasis for medical research. When the growth of health-care expenditures seemed virtually unlimited, researchers were primarily interested in finding new drugs, new treatments, or new tests that made a positive contribution to health regardless of the benefit–cost ratio. All that was needed was a demonstration that the innovation was better than existing alternatives. Under the competitive, cost-conscious conditions that are likely to prevail in the 1980s, researchers will be expected to answer the questions "How much better is the innovation?" and "How do the expected benefits compare with those that could be obtained if the resources were used in some alternative way?"

Finally, the new modes of practice and the new demands on research will inevitably have an impact on the form and content of medical education. For instance, the training of interns and residents will increasingly incorporate such questions as "Why *did* you order this test?" as well as "Why *didn't* you. . . ?" At the undergraduate level the scientific base will probably be broadened to incorporate elements of the behavioral, social, and decision sciences as well as the traditional biological sciences. Adaptation will be particularly difficult because the development of new programs, new courses, and new techniques must occur in what will be virtually a no-growth situation.

Thus the coming challenge will require skillful administration and considerable understanding and cooperation within and between the "gown" and "town" medical communities. In the past 30 years physicians have been the beneficiaries of historical developments that were mostly beyond their control; in the decade ahead they are likely to be the victims. Their responsibility, as always, is to do the best they can under the given circumstances.

## ACKNOWLEDGMENT

I am happy to acknowledge the financial support of the Robert Wood Johnson Foundation.

## REFERENCES

1. Hughes, E. F. X., Lewit, E. M., and Lorenzo, F. V., 1975, Time utilization of a population of general surgeons in community practice, *Surgery* **77**:371–383.
2. Enthoven, A. C., 1980, *Health Plan: The Only Practical Solution*, Addison-Wesley, Reading, Mass., p. 63.

# Index

U.C.W. ABERYSTWYTH